Principles of

HUMAN PHYSIOLOGY

SECOND EDITION

Lab Manual

Geoffrey Goellner
Rachel Cohen
Penny Knoblich
Tanya Simms

Kendall Hunt
publishing company

BIOPAC BSL 4.0 screenshots courtesy of BIOPAC Systems, Inc.

Cover image © Shutterstock, Inc.

Kendall Hunt
publishing company

www.kendallhunt.com
Send all inquiries to:
4050 Westmark Drive
Dubuque, IA 52004-1840

Copyright © 2015 by Kendall Hunt Publishing Company

ISBN 978-1-5249-2245-0

Published in the United States of America

CONTENTS

Homeostasis

Study Questions

1. Why is homeostasis considered a dynamic state, and not an absolute state?
2. What are the various components of a typical negative feedback loop?
3. How does negative feedback act to maintain homeostasis?
4. How is each of the following determined: setpoint, range, and sensitivity?
5. What happens when a physiological setpoint is *reset*? Give an example of this.
6. Why might resetting a setpoint be beneficial?
7. What is a positive feedback system (loop)? Give an example.
8. Why does a positive feedback loop that is not stopped result in death?
9. What is feed forward? Give an example of this. What advantage does a feed-forward system in homeostasis have when compared to a negative feedback system?
10. In the negative feedback control of heart rate, heart rate is not the regulated variable. What is the body regulating?
11. Heart rate in the absence of any nervous system influence is 100 beats per minute. Keeping this in mind, why do we need both a sympathetic and parasympathetic nervous system in heart rate regulation?

Applied Questions (Answers in Back)

1. Plasma level of thyroid hormone (TH) is controlled by negative feedback. The release of TH is stimulated by another hormone from the pituitary gland known as TSH. The loop is as follows:

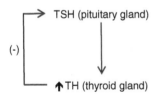

A patient with hypothyroidism ($\downarrow$ TH) due to a defective thyroid gland would likely have the following:
   a. A very high TSH.
   b. A very low TSH.
Explain your choice.

2. Jill Jones is sick with Strep. She has a very high temperature and is shivering.
   a. Why is she shivering instead of sweating?
   b. Has her negative feedback system for temperature regulation failed?

3. Just as we begin to exercise, <u>before</u> our body's needs have increased, our respiration and heart rate get faster. What type of mechanism is this?

Introduction

As macromolecular (Ex. protein) structures have evolved to function under very specific conditions, human cells are dependent upon a relatively constant internal environment (any fluid in immediate contact with a cell in the body is referred to as the **internal environment**). Thus, the internal environment of the body needs to be regulated to maintain the composition, temperature, and volume of extracellular fluid surrounding our cells. The term **homeostasis** is used to refer to this very important "stable state" of the internal environment surrounding our cells.

Homeostasis is usually maintained by **negative feedback loops** in which any change in a specific condition of the internal environment (**physiological variable**) is returned (corrected) to its normal value (**setpoint**). Because a physiological variable (regulated variable) is permitted for a short time to vary above or below the setpoint, homeostasis should be considered a **dynamic state**, not an **absolute state**. Furthermore, on occasion a physiological setpoint may be reset to a different value- typically only for a short time. An example of this would be when the body induces a fever (elevated body temperature) to help fight off an infection. A negative feedback system begins with a change in a physiological variable (**error signal** or **stimulus**) that is sensed by a **receptor** (typically a protein that changes shape in response to the stimulus). The information is then transferred to the **integrating center** (typically the brain or spinal cord) via a sensory nerve (**afferent pathway**). The integrating center then orchestrates an appropriate response by sending information (via the **efferent pathway**) to an **effector** organ to have the desired **response -** returning the physiological variable back toward its setpoint value. A flow chart summarizing these ideas (using temperature as an example) is shown below:

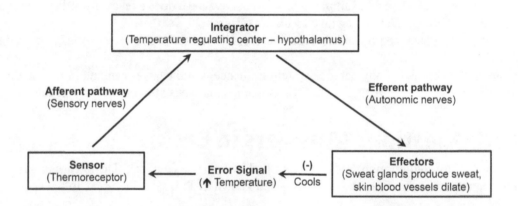

Feed-forward processes are physiological or behavioral responses designed to anticipate a change in a physiological variable, and make the response even before the "change" occurs. Thus, while negative feedback responses are **reactive**, feed-forward processes are **proactive**. Feed-forward processes can *prevent* a change in a physiological variable, or can decrease the magnitude of the change from the setpoint (keeping the variable even closer to the setpoint). Examples of feed forward include: increased heart and breathing rates *before* an athletic event begins, increased salivary and stomach secretions *before* consuming a meal.

Occasionally, the body would like to move a certain condition AWAY from a physiological setpoint. This can be done via a **positive feedback** loop in which the response proceeds in the *same direction* as the original change. Thus, in contrast to negative feedback loops, positive feedback is designed to ***not*** maintain homeostasis, but instead enhance the change even further from the original setpoint. Typically, this "burst away" from a setpoint is limited in duration, and must be terminated to interrupt the cycle. An example of a beneficial positive feedback loop would be estrogen secretion that leads to ovulation: ↑ estrogen →↑ LH →↑ estrogen →↑ LH, etc. Positive feedback loops are also used to generate action potentials that underlie neurotransmission: ↑sodium entry →↑open sodium channels →↑sodium entry →↑open sodium channels, etc. Unfortunately, "unchecked" harmful positive feedback loops can also occur (Ex. heat stroke, heart failure) that inexorably move the variable further and further from the setpoint, until death occurs.

Today's lab will demonstrate some of these ideas, by measuring your heart rate to observe how negative feedback is used to maintain blood pressure homeostasis (significant blood pressure is always required within the body to ensure adequate blood flow to all vital organs). Thus, <u>blood pressure</u> is the regulated variable, and heart rate will be adjusted to maintain a normal blood pressure. An increase in heart rate will cause an increase in blood pressure, while a decrease in heart rate will lead to a decreased blood pressure. Luckily, we do not have to consciously be aware of either our heart rate or blood pressure, as the **autonomic nervous system (ANS)** takes care of this for us. The **sympathetic division** of the ANS "automatically" increases heart rate when blood pressure drops below the setpoint value, while the **parasympathetic division** decreases heart rate when blood pressure increases above the setpoint value.

Setpoint will be calculated by finding the mean of the measured heart rate values. The difference between the average error signal above and below the setpoint [(average of above setpoint values) – (average of below setpoint values)] is referred to as the **range** of a negative feedback process. The following figure shows typical data points of heart rate plotted over time (dashed line is the set point).

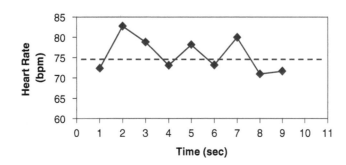

Experiment: Negative Feedback Control of Heart Rate
Program Setup

1. Log in to the computer at your station
2. Turn the MP36 on using the switch at the back of the unit.
   a. The pulse transducer should be plugged into **channel 1**.
3. Double click on the *BIOPAC Student Lab 4.1* icon on the desktop

Note: If you get the message shown in **Figure 1.1**, choose **OK** and then **Quit** when the dialog box comes up. Turn the MP36 unit on and repeat step 3.

Figure 1.1

4. The dialog box below (**Figure 1.2**) should appear. If no lessons are listed then your MP36 unit is not turned on. Turn your unit on, close the program, and open it again.

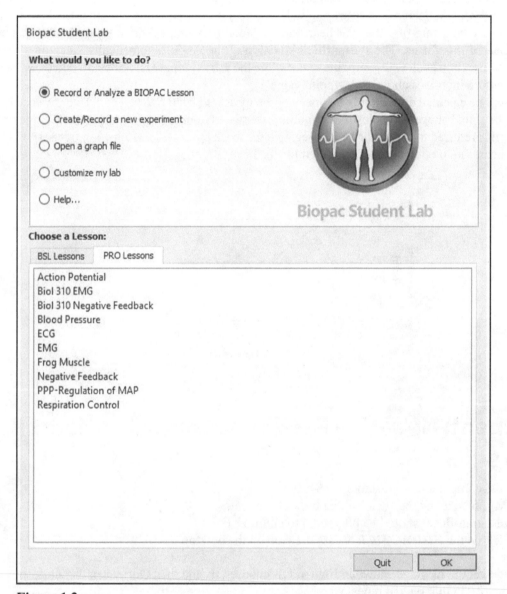

Figure 1.2

5. Click on the PRO Lessons tab, select Negative Feedback and then click OK.
6. The screen shown in **Figure 1.3** should appear. Two channels, channel 1, which records the PPP (pulse) wave and channel 40 that records pulse rate by calculating it from channel 1, should be apparent.

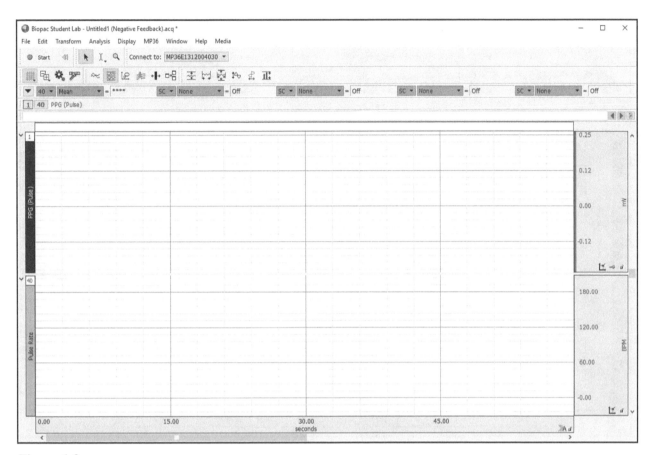

Figure 1.3

Subject Setup

1. Attach the pulse transducer snugly to any finger except for the thumb using the expandable black strap. The wire of the pulse transducer should come off the fingertip (*i.e.*, it should be pointed away from the body).
2. The subject should sit with his/her arms resting on the table and feet on the floor.
3. The subject must minimize his/her hand and finger movements, which will alter the recording.

Calibration

1. Click on Start in the upper left corner of the graph window. Record the subject's pulse pressure for 15 seconds (watch the time scale on the bottom of the graph window) and then click on Stop. **DO NOT REMOVE THE PULSE TRANSDUCER!!**
2. To autoscale the data, choose Autoscale Waveforms from the Display menu or right click on the graph window and select Autoscale Waveforms from the menu that appears.
3. If your data does not resemble **Figure 1.4** contact your instructor.

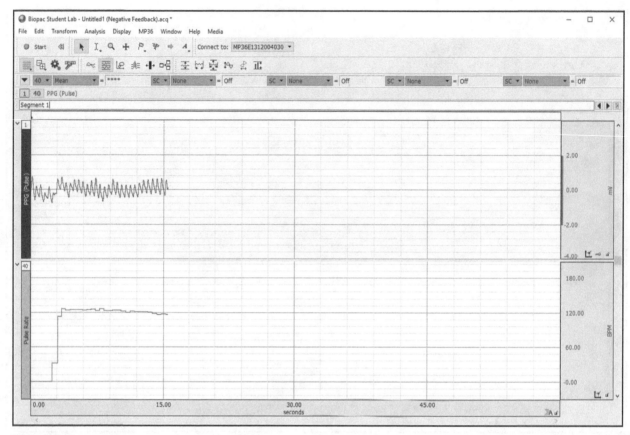

Figure 1.4

Saving Your Data

1. Under the File menu click on Save as.
2. In the *Save graph as?* screen that appears click on Desktop (in the left hand column) and then on Student Data.
3. Double click on your section's folder.
4. Type in **Negative Feedback** as the file name.
5. Click on Save.

Data Collection

1. With the subject sitting relaxed and still, click on Start.
2. You will get a message asking if you should overwrite existing data. Select Yes.
3. Collect data until the computer stops on its own (~240 seconds total).
4. Save your data (File>Save).

Data Analysis

I. Measuring Average Heart Rate over Time

1. Click on the second icon from the left under the Start menu and then on journal in the drop down menu that appears (See **Figure 1.5**). A journal window will appear at the bottom of the graph window.

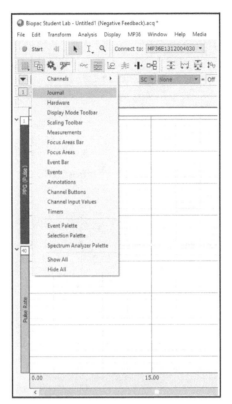

Figure 1.5

2. Select the I–icon (the box next to the magnifying glass) to the right of the Start button at the top of the graph window.
3. Scroll back to the beginning of the data by using the scroll bar at the bottom of the graph window.
4. At the 15 second mark, use the cursor to drag across the second 15-second period, *i.e.*, from 15–30 seconds on the time scale.
5. Click Control-M (Ctrl-M). The mean heart rate for that 15 second interval should appear in the journal. **Note:** If letters or labels appear in the journal along with the numbers, contact your instructor.
6. Repeat measurement of alternate 15-second intervals until the end of the data collection period. You should have a total of eight measurements when you are finished.
7. Save your data (File>Save).

II. Graphing the Heart Rate

1. Copy your journal data by dragging over them and then pressing Control-C (Ctrl-C) on the keyboard.
2. Open the Excel Templates folder within the Bio 330 folder on the desktop.
3. Double click on the Negative Feedback template to open.
4. In the Negative Feedback workbook, select the cell directly under the cell labelled Heart rate.
5. Paste your data into the sheet by pressing Control-V (Ctrl-V) on the keyboard. A heart rate line should appear on the graph as shown in **Figure 1.6**.

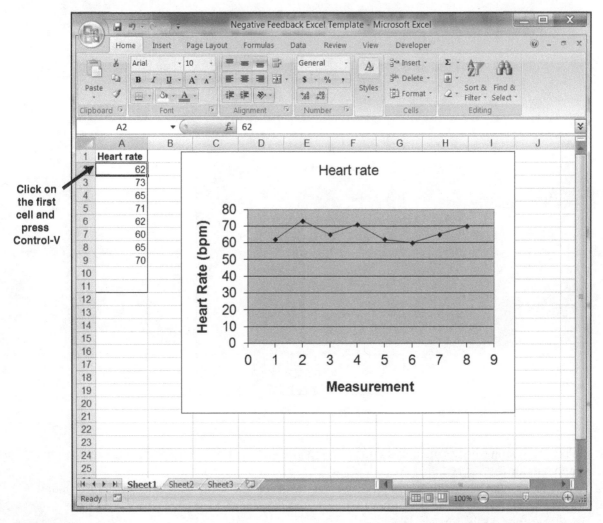

Figure 1.6

6. Be sure that your *y*-axis scale (heart rate) is set to read between 0 (minimum) to ~100 (maximum).
7. Print your Excel worksheet to the printer labelled **TS380 - HP1505N on MAVPRINT**. To do so, click any cell on the sheet (do not select the graph, or your data will not print) and then on File > Print. Make sure that the orientation is Landscape and the Fit Sheet on One page box is selected.
8. Once printed, exit Excel without saving your graph. Close the MP36 graph window and turn the Biopac off via the switch in back of the MP36 unit.

Homeostasis Data Sheet

Group Number: _____   Name: _____

Calculate the following (show your work):

a. Your heart rate set point (mean or average).

b. Your heart rate range (average of values above the setpoint) – (average of values below the setpoint).

c. Your heart rate sensitivity (range ÷ 2).

Using your knowledge of negative feedback, explain why the values you measured are not all the same. What is happening physiologically to produce these different values?

Diffusion & Osmosis

Study Questions

1. Define diffusion. What factors affect the rate of diffusion?
2. What molecules can cross the cell membrane using simple diffusion? What factors affect the rate of simple diffusion across the membrane?
3. Does simple diffusion occur from a *higher* to *lower* concentration, or from a *lower* to *higher* concentration?
4. Define facilitated diffusion.
    a. What factors affect the rate of facilitated diffusion?
    b. Does facilitated diffusion require energy? Why or why not?
5. What are the two forces that act on an ion to determine its <u>net force</u>?
    a. How does this relate to the rate of diffusion across the cell membrane?
6. What is permeability? How does it relate to diffusion across the cell membrane?
7. How is passive transport different from active transport? Give an example of each.
8. Define osmosis.
9. Does water move from lower to higher <u>solute</u> concentration, or from higher to lower <u>solute</u> concentration? Explain using the *concentration* of water versus the concentration of solute.
10. What is one mole? How is this related to Avogadro's number?
11. In determining osmosis, why is concentration converted to moles/liter from grams/liter?
12. Why is a solution concentration converted to osmolarity when determining osmosis?
13. Why does a 1.0 molar solution of glucose have a different osmotic effect than a 1.0 molar solution of sodium chloride?
14. Using the concentration of water, explain:
    a. Why a cell *shrinks* when placed in a hypertonic salt solution?
    b. Why a cell *swells* when placed in a hypotonic salt solution?
    c. Why a cell does not change in volume when placed in an isotonic salt solution?
15. When is a solution **isotonic**, **hypertonic**, or **hypotonic**?
16. What happens to a cell with an osmolarity of 300 mOsm, when it is placed into solution with an osmolarity of 400? Shrink, swell, or nothing?
17. Explain why the water level climbed up into the thistle tube in our lab demonstration.

Applied Questions (Answers in Back)

1. Perry pathologist just placed a slide of cells under the microscope. He discovered that all the cells appear swollen, and some have burst. He must have washed the cells with the wrong solution. Was the solution hypertonic, hypotonic, or isotonic? Explain.
2. Carrie's boss has told her to mix up an isotonic (300 mOsm) salt solution to rinse blood cells with. Carrie cannot remember how many grams of salt to add to one liter of water. Can you help her?

Introduction

Cellular plasma membranes are composed primarily of proteins and amphipathic lipids. These macromolecules naturally form bilayers that arrange in such a way as to form a **significant hydrophobic (water fearing) region** in the interior aspect of the bilayer (hydrophobic tails), and two hydrophilic (water loving) regions on the exterior aspect of each bilayer (polar head groups). This precise arrangement results in a bilayer that is **selectively permeable** to the transit of different types of substances (based primarily on molecular size and hydrophobicity). Thus, certain molecules can easily cross cellular membranes unaided- while others require "help." Indeed, there are a number of different mechanisms used to move various types of substances into or out of the cell.

Diffusion refers to the <u>random</u> movement of molecules ultimately resulting in an equal distribution (concentration) of that molecule on both sides of the membrane. Diffusion occurs naturally due to the kinetic energy of molecules, and always results in the **net flow** of the molecule **DOWN** its concentration gradient (the concentration gradient refers to the difference in concentration of the molecules between two locations- Ex. inside versus outside of the cell). During diffusion then, the net flow of molecules moves from an area of higher to lower concentration via a **passive process** that **does not** require **energy**. Diffusion rates are affected by various parameters such as temperature, magnitude of concentration gradient, and permeability (ease with which a substrate can cross the membrane). It is important to note that there are **different types of diffusion** which allow cells to move different types of molecules across their cellular membranes.

Simple diffusion allows the passage of **hydrophobic** (lipid-soluble) substances directly across the hydrophobic interior of cellular membranes. As this a form of diffusion, it is important to emphasize that **no energy is required**, and molecules move **down** their concentration gradient (from high to low) directly across the membrane completely unaided. Not many substances can manage to do this in the cell; however, blood gases (O_2, CO_2) and steroid molecules can for example. The rate of simple diffusion is dependent upon: relative hydrophobicity of permeating molecule, size of the permeating molecule, concentration gradient magnitude, temperature, and thickness/surface area of the membrane.

Facilitated diffusion uses some form of **transport protein** to "facilitate/help" mediate the movement of **hydrophilic** (water-soluble) substances across the hydrophobic interior of the lipid bilayer. Indeed, as most cellular molecules are hydrophilic, a large number of different molecules move across cellular membranes in this manner. Importantly, as this is a version of diffusion, substances still move from high concentration to low (down gradient) with no energy required.

An example of facilitated diffusion includes the movement of glucose across the cell membrane via a glucose transporter. Glucose is a relatively large hydrophilic molecule, and therefore requires "help" to cross the hydrophobic interior of the plasma membrane bilayer. The glucose transport protein "facilitates" the movement of glucose either into or out of the cell (depending on the concentration gradient) by binding specifically to glucose and helping it to cross the plasma membrane. Although there is a protein involved (glucose transporter) that changes shape/configuration upon glucose transit- there is still no energy required as glucose moves DOWN its concentration gradient. Please note, the glucose transporter is very **specific** for what it transports (glucose only), and can **saturate** (has a finite number of glucoses that it can move per second).

Ion channels are another example of facilitated transporters (no energy required). Ion channels are proteins embedded in the plasma membrane that, when open, allow the movement of ions across the bilayer. Even though ions are quite small, they still can't cross the plasma membrane unaided because they are too hydrophilic; indeed, they are **charged**. Ion channels are very **specific** for which ions they let through (Ex. sodium channels only allows Na^+ transit), and typically need to be "told" to when to open (we call this **"gating"**). Different channels are gated open in different ways (Ex. ligand binding, mechanical distortion, or electrical changes). The direction in which an ion will flow through an open channel is dictated by two main factors- electrical and chemical force. Since ions contain an inherent electrical charge they tend to be drawn towards the opposite charge (remember "opposites attract" in electricity). Also, ions are subject to concentration gradients (just like any other molecule). Thus, two forces (electrical and chemical) act on ions to dictate direction of flow through an open channel- we call this the **electrochemical gradient**.

Active transport allows the movement (pumping) of molecules across membranes **against or up** their concentration gradient (lower to higher concentration). This always requires a transport protein, and also requires **energy (Ex. ATP)**. An example of active transport is the movement of sodium and potassium ions across the plasma membrane **against** there normal gradient via the protein - Na$^+$/K$^+$ ATPase. Because of their respective electrochemical potentials, sodium would typically like to come into the cell and potassium would typically like to leave. However, using energy, the Na$^+$/K$^+$ ATPase instead "pumps" these ions against their normal gradient - causing sodium to leave and potassium to enter the cell. Indeed, in this way, the Na$^+$/K$^+$ ATPase **creates** an unequal distribution of ions (lots of sodium outside and lots of potassium inside of the cell). This unequal distribution of ions can then be used by certain cells (Ex. neurons) to transmit information in the body.

Osmosis refers specifically to the movement of **water** across biological membranes. Essentially, water moves via facilitated diffusion - as most water crosses the membrane via special water channels called **aquaporins.** Assuming the aquaporin is open, water behaves just like every other molecule in that it will move down its concentration gradient. Please note, however, that as water moves down its own concentration gradient it will be moving to an area of higher <u>solute</u> concentration. Thus, osmosis can be defined as the <u>facilitated diffusion</u> of water across a selectively permeable membrane, from a region of <u>higher water concentration</u> (lower solute concentration) to a region of <u>lower water concentration</u> (higher solute concentration). The rate of water movement will be affected by both the <u>concentration difference</u> of water molecules across the cell membrane, and the <u>number of open aquaporins</u>.

Finally, there are a number of important terms describing solutions that should be defined prior to lab. The term **osmolarity** refers to the total number of solute *particles* in a solution. Thus, a 1.0 molar solution of a <u>nondissociating</u> substance, such as <u>glucose,</u> will contain Avogadro's number of molecules (6.02×10^{23}); so a <u>1.0 molar</u> solution of **glucose** will produce a 1.0 Osmolar solution. On the other hand, molecules of a <u>dissociating</u> substance such as sodium chloride (salt) will split into <u>two particles</u> when placed into water (Na$^+$ and Cl$^-$); so a <u>1.0 molar</u> solution of NaCl will produce a <u>2.0 Osmolar</u> solution. Importantly then, a 1M solution of NaCl will have <u>double</u> the osmotic effect on a cell as compared to the 1.0 molar solution of glucose. Because the concentration of solutes in normal cells and body fluids is low, the term **milliosmolar** is typically used instead of **osmolar** (remember a milliosmole is 1/1000 of an osmole). **Tonicity** is another physiological term commonly used to describe solutions, and is important in regard to effect on cell volume. An **isotonic solution** is a solution that has the <u>same</u> osmolarity when compared to that of a cell (~300 mOsm). Thus, the ultimate <u>water concentration is the same</u> in both the solution as well as in the cell, and there is no change in cell volume via osmosis when a cell is placed into this solution. A **hypotonic solution** is a solution that contains a <u>lower</u> osmolarity when compared to a cell. Thus, the <u>water concentration is higher</u> in the solution than in the cell, and when placed into this solution the cell swells as water enters by osmosis. A **hypertonic solution** is a solution that has a <u>higher</u> osmolarity when compared to a cell. Thus, the <u>water concentration is lower</u> in the solution than compared to the cell. When placed into a hypertonic solution, a cell shrinks as water leaves by osmosis.

Osmosis

Definitions and Aid to Calculations

1. <u>Mole</u>: The amount of a substance that is equal to its molecular weight in grams.
   - One mole of any substance weighs the molecular weight in grams.
   - One mole of any substance has the same number of molecules (Avogadro's number).
   - Therefore, a mole of NaCl and a mole of glucose have the same number of molecules, but do not weigh the same (NaCl = 58.5 grams; glucose = 180 grams).
2. <u>Molar</u>: 1.0 Mole of a substance dissolved in 1000 milliliters (one liter) of a solution.
3. <u>Osmolarity</u>: The sum of all the solute particles of a substance, once it is in solution.
   - The osmolarity of a one molar solution of glucose = 1 Osm.
   - The osmolarity of a one molar solution of NaCl = 2 Osm (One mole of Na$^+$ plus one mole of Cl$^-$ = 2)
4. Milliosmolarity = osmolarity $\times$ 1000.
5. Molecular weights:
   - NaCl = 58.5 g/mole
   - Glucose = 180 g/mole
6. dl = deciliter = 100 ml (10 dl = 1 liter).

To solve the calculations, use the following steps:

a. Convert g/dl to g/l by multiplying the grams times 10

$$\frac{g}{dl} \times \frac{10\,dl}{liter} = \frac{g}{liter}$$

b. Convert grams to moles: divide grams by the molecular weight of the substance

$$\frac{g\ NaCl}{liter} \times \frac{mole}{58.5\ g\ NaCl} = \frac{moles\ NaCl}{liter}$$

c. Convert moles to osmoles

$$\frac{moles\ NaCl}{liter} \times \frac{2\ osmoles\ NaCl}{mole\ NaCl} = \frac{osmoles\ NaCl}{liter}$$

d. Convert osmoles to milliosmoles

$$\frac{osmoles\ NaCl}{liter} \times \frac{1000\ milliosmiles}{osmole} = \frac{milliosmoles}{liter} = milliosmolarity$$

Example: 4 g NaCl/dl

a. $\dfrac{4\,g}{dl} \times \dfrac{10\,dl}{liter} = \dfrac{40\,g}{liter}$

b. $\dfrac{40\ g\ NaCl}{liter} \times \dfrac{mole}{58.5\ g\ NaCl} = \dfrac{0.684\ moles\ NaCl}{liter}$

c. $\dfrac{0.684\ moles\ NaCl}{liter} \times \dfrac{2\ osmoles\ NaCl}{mole\ NaCl} = \dfrac{1.368\ osmoles\ NaCl}{liter}$

d. $\dfrac{1.368\ osmoles\ NaCl}{liter} \times \dfrac{1000\ milliosmoles}{osmole} = \dfrac{1368\ milliosmoles}{liter} = 1368\ mOsm$

You may run the conversions together:

$$\frac{14\,g}{dl} \times \frac{10\,dl}{liter} \times \frac{mole}{58.5\,g} \times \frac{2\ osmoles\ NaCl}{mole\ NaCl} \times \frac{1000\ milliosmoles}{Osmole}$$

Osmosis Demonstration

A thistle tube containing a concentrated sugar solution (molasses) is placed into a beaker of distilled water. The level of molasses in the tube is demarcated by your instructor at the beginning of class.

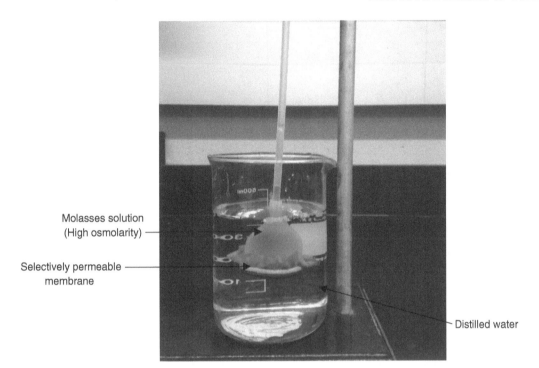

Molasses solution (High osmolarity)

Selectively permeable membrane

Distilled water

Osmolarity Case Study

You are a nurse at the Mayo Clinic emergency room where there was a mix-up with bags of different I.V. fluids. You suspect that three patients may have received fluid in their IV that was not of the correct osmolarity. To be able to remedy this problem, you take blood samples from each of the three patients, which you will use to compare to a normal blood sample under the microscope. Draw your observations in the space provided in the data sheet and use your background knowledge of osmosis to determine whether each patient received hypotonic, hypertonic or isotonic fluid in their IV.

Diffusion and Osmosis Data Sheet

Group Number: _____   Name: _____

Thistle Tube Demonstration

1. Where is the *higher* water concentration, inside the thistle tube or outside in the beaker?

2. Will water tend to enter or leave the tube by osmosis?

3. Measure the height of the liquid level in the tube (beginning of class): _____

4. Predict what you expect the height to be at the end of class: _____

5. Measure the actual height (at the end of the class period): _____

Osmosis Calculations

Complete the chart below. (Answers in back)

Solute (g/dl)	Molarity (moles/L)	Osmolarity (Osm)	Milliosmolarity (mOsm)	Effect on a cell (300 mOsm)	Tonicity of solution
4 g NaCl/dl	**0.684**	**1.368**	**1368**	**Shrink**	**hypertonic**
3 g NaCl/dl					
3 g glucose/dl					
0.5 g NaCl/dl					

A solution of NaCl has a milliosmolarity of 200. How many grams of NaCl were added to a **1 liter** container of pure water to produce this solution? (Answer in back)

<u>Osmolarity Case Study</u>

	Drawing	Diagnosis (Explain why)
Normal Blood		N/A
Patient 1		
Patient 2		
Patient 3		

Action Potential

Study Questions

1. How is the concentration of Na^+ inside the cell different from the concentration outside the cell? What about K^+?
2. What maintains this unequal distribution of ions?
3. What is a membrane potential? What does the sign that precedes the number mean?
4. What is an **action potential**?
5. Fill in the ion movements through each numbered phase of the following recording of a monophasic neuron action potential.

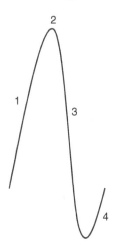

6. Mark where each voltage-gated channel begins to open, close, or inactivate.
7. Since each action potential in a single neuron is the same amplitude, explain how strength of stimulus is encoded in the central nervous system.
8. What is a **refractory period**? What is the ion channel and gate involved?
9. Explain in terms of ion channels and gates the difference between the absolute and relative refractory periods.
10. How is the electrode placement in our frog nerve (biphasic compound action potential or CAP) different than when studying a single neuron (monophasic action potential)?
11. What causes a stimulus artifact to appear on the action potential recording channel?
12. Why does higher stimulus voltage increase the amplitude of the CAP in our sciatic nerve experiment, but not in a single nerve?
13. The following is a picture of a compound biphasic action potential:
    a. At which letter is the first electrode depolarizing?

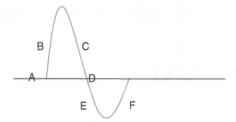

b. At which letter is the first electrode repolarizing?
c. At which letter is the depolarization between the two electrodes?
d. At which letter is the second electrode depolarizing?
e. At which letter is the second electrode repolarizing?
14. Define threshold voltage: How did we determine this?
15. Define maximal voltage. How did we find this?
16. How did we measure absolute and relative refractory periods in our frog nerve?
17. How did we determine the speed of conduction in our frog nerve?

Applied Questions (Answers in Back)

1. The sensory nerves carry information from the skin to the brain. John touched his sister's arm to get her attention. She ignored him, so he punched her arm. From any single sensory neuron from the affected area, compare the frequency of action potentials from the touch to that of the punch.
2. Certain medications may make the resting membrane potential of a single neuron more negative. How might this affect the stimulus voltage that would need to be applied to bring the neuron to threshold (larger, smaller, unchanged)? Would this affect the size of the action potential?

Introduction

Neurons typically communicate with other cells (such as another downstream neuron or a muscle cell) using electrical/chemical signals. The electrical part of this communication is referred to as the **action potential**. Action potentials are caused by small fluxes of ions (Na^+ and K^+) across the axonal plasma membrane of neurons. The unequal ion distribution that exists across a "resting" neuron's plasma membranes underlies how action potentials occur. It is important to note that there is typically a higher concentration of K^+ inside a resting neuron than outside, while Na^+ has a higher concentration outside the resting cell than inside. This unequal distribution of ions is produced and maintained predominantly by the $3Na^+/2K^+$ ATPase pump that continually moves Na^+ out of the cell, and pumps K^+ into the cell (using ATP). It is also important to note, that as the Na^+/K^+ ATPase pumps 3 positively charged ions out (Na^+) for every 2 positive ions into the cell (K^+), the Na^+/K^+ ATPase is considered "electrogenic;" it creates an unequal distribution of charge across the membrane too (inside of resting neurons is negative relative to outside the neuron). This unequal distribution of charge across the plasma membrane is called the **membrane potential**, and in most cells measures around -70mV (the sign in front of the membrane potential value indicates if the inside of the cell is negative or positive). Thus, there is both an unequal distribution of ions (Na^+ and K^+) and charge across resting neuronal plasma membranes.

During an action potential, neurons utilize small changes in membrane potential to help communicate to other cells. As noted above, at rest the inside of a neuron has an excess of negative charges when compared to that of the outside. During an action potential, for a brief moment, the inside of the neuron gains positive charge (Na^+ enters through open channels), and becomes positive when compared to the outside of the cell. This rapid entry of Na^+ into the neuron is considered the **depolarization phase** of the action potential. Shortly thereafter, Na^+ channels close and K^+ channels open- leading to a massive efflux of positively charged K^+ ions out of the cell (**repolarization phase**). K^+ efflux causes the inside of the neuron to become negative again, and repolarize near to resting membrane potential (-70mV). These transient (msec) changes in membrane potential (action potential) ultimately signal events to occur in the synapse that lead to "neurotransmission" of information (release of neurotransmitter) between the neuron and a downstream postsynaptic cell.

The typical action potential discussed in the textbook, lecture, and above is called a **monophasic action potential**. The monophasic action potential is recorded from a <u>single</u> neuron axon, and during recording, one electrode is placed *inside* the neuron while the other electrode is placed on the *outside* of the cell. The voltage difference across the membrane (inside vs outside) is then recorded over time.

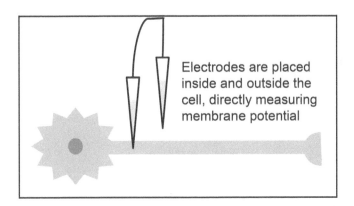

A monophasic action potential recorded from a <u>single</u> neuron would look like the following diagram:

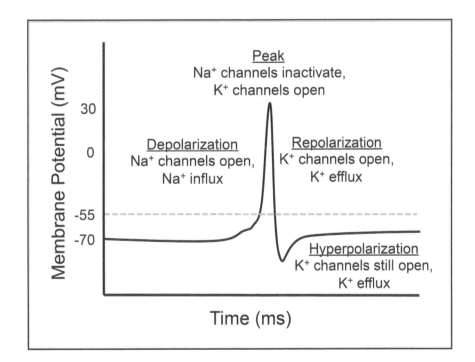

The action potential from a <u>single</u> neuron is an **all-or-none** phenomenon, and does not increase in amplitude (height) when the strength of stimulus is increased over and above the threshold voltage (suprathreshold stimulus). Thus, a larger stimulus cannot create a larger action potential. However, it can stimulate a <u>series</u> of action potentials. Stimulus strength can, therefore, be encoded by the frequency of action potentials being generated in a single neuronal axon (**frequency coding**).

The action potentials that will be measured in today's lab are called **biphasic or Compound Action Potentials (CAP)**. The biphasic action potential is the result of voltage changes recorded when **both** measuring electrodes are placed **<u>outside the cell</u>**, one further along the nerve axons than the other. The recording represents the difference in voltage between the two electrodes, **not** the difference between *inside* the cell and *outside*. At resting potential, the two electrodes will have the <u>same</u> voltage (no difference between them), and the recorded resting voltage will read <u>0 mV</u> (rather than −70 mV).

The frog sciatic nerves used in today's experiment have been dissected and prepared by your lab instructor. Please note, that the sciatic nerve is a compound nerve (consisting of both motor and sensory axons). This "bundle of axons" emanates from the spinal cord and innervates the frog's leg.

Since the dissected nerve is now isolated from the receptors in the frog's body that normally generate the action potentials, an "outside" electrical stimulus will be used to bring the axons to threshold. After dissection, the nerve was bathed in **Ringer's** solution before it was placed on the electrode wires. Frog Ringer's solution is similar to the ion composition of the extracellular fluids normally found in a living frog (please see chart below).

Ringer's Solution:	
Sodium chloride	6.000 grams
Potassium chloride	0.075 grams
Calcium chloride	0.100 grams
Sodium bicarbonate	0.100 grams
Double distilled water to make 1 Liter	

The Compound Action Potential

The following section describes the movement of the action potential down the nerve axons, and the resultant changes in the measuring electrodes and computer screen. At rest, the inside of the cell is negative with respect to outside, so positive charges (attracted by the negative cell) line up on the outside of the neuron. Since both recording electrodes are on the <u>outside</u> of the cell, both are surrounded by positive charges, and no electrical difference exists. The computer recording is 0 mV, which means there is <u>no difference</u> in charge between the two electrodes.

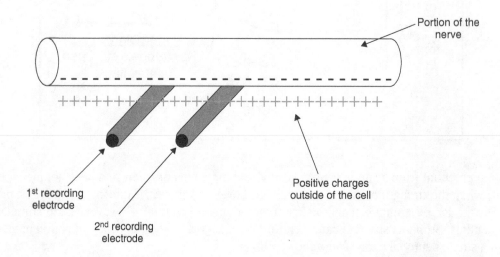

An action potential is generated by using stimulating electrodes to the left of the recording electrodes. Depolarization (inside now *positive*) of the axon membrane at the first electrode causes negative charges to become attracted to the outside of the membrane. Thus, the first electrode is now surrounded by <u>negative</u> charges. A charge difference (negative vs. positive) now exists between the <u>first</u> and <u>second</u> recording electrode, causing the computer recording to move *upward*. The depolarization phase spreads past the membrane of the first electrode, and the membrane at the first electrode **repolarizes**. As the inside of the cell returns to its negative state, the first electrode is again surrounded by *positive* charges. As the electrical difference between the electrodes disappears (both once again surrounded by positive charges), the computer recording returns to zero. At zero, the depolarization phase is now <u>between the two recording electrodes</u>. Depolarization of the axon membrane then reaches the second recording electrode. The charge difference between the **first** and **second** electrode causes the computer recording to move downward. Note that the direction is *reversed* from the earlier change, because the charge difference is *reverse* of what occurred during depolarization at the first electrode. Eventually, the depolarization spreads past the membrane of the second electrode, and the electrical difference between the electrodes disappears, causing the computer recording to return to zero.

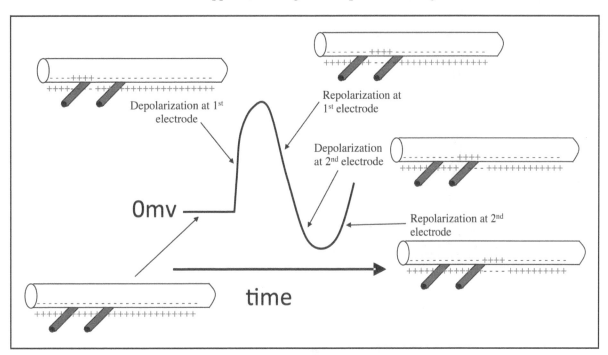

The Frog Sciatic Nerve: A Bundle of Neuron Axons

The frog sciatic nerve used in this lab consists of many independent neuron axons (bundled together), and is therefore called a <u>compound nerve</u>.

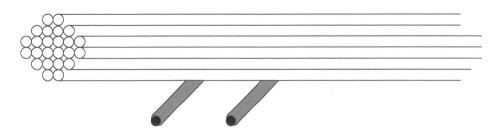

The independent neuronal axons are of many different types and sizes (sensory, motor, etc.), and therefore the <u>threshold</u> for stimulation of an action potential will vary among different neuronal types. A small stimulus will cause action potentials only in neuron axons with <u>low threshold</u> voltages, and a large stimulus will create action potentials in axons with <u>both</u> low and high threshold voltages. Simply put, the <u>larger</u> the stimulus voltage - the <u>greater the number</u> of individual neuronal axons that will be stimulated to have an action potential. The computer recording ***adds up*** all of the independent depolarizations in each individual axon. This produces a CAP whose amplitude (height) is directly related to the number of individual axons generating action potentials. Thus, a low stimulus voltage will produce a small CAP visible on the computer screen, and a larger stimulus voltage will produce a larger CAP. Please note, however, that each individual neuron still has an all-or-none action potential, but the computer recording varies in amplitude because it "adds up" all of the separate action potentials. Thus, in our recording of a compound nerve, we will observe a CAP amplitude that varies with the strength of stimulus voltage. This principle is illustrated in the following diagrams.

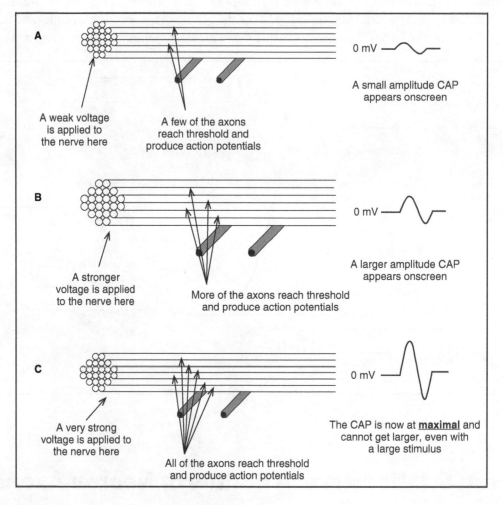

In today's lab, you will perform three experiments.

Experiment 1: Determining Threshold and Maximal Voltage

The sciatic nerve is made up of many neuron axons that have different thresholds for stimulating an action potential. At smaller voltages, only those axons with low thresholds will be stimulated to have action potentials. As the stimulus voltage is increased, more axons reach threshold, and more axons have all-or-none action potentials. The action potentials are added together by the computer to produce a greater CAP amplitude.

In this experiment, you will determine both the threshold and the maximal voltage for your nerve. The threshold voltage is the lowest stimulus voltage required to bring enough axons to threshold to create a visible deflection (small CAP) on the computer screen. To determine the threshold voltage, you will stimulate your nerve with a low stimulus voltage and gradually increase the voltage until a response is visible. Maximal voltage, on the other hand, is the lowest voltage that brings all the axons in the nerve to threshold. Once you have determined the threshold voltage, you will continue to increase the voltage until the last CAP is the same size as the one before it. It should be noted that applying a supramaximal voltage will <u>not</u> increase the amplitude of the CAP because once maximal voltage has been achieved, all axons in the nerve are already having action potentials.

Experiment 2: Determining the Speed of Conduction of the Action Potential Down the Nerve

Speed can be determined by measuring the time a moving object takes to traverse a known distance.

In this experiment, the distance between electrodes on the nerve is known and the time for the action potential to travel from one set of measuring electrodes (channel 2) to the next set of measuring electrodes (channel 3) can be determined. By taking the known distance (5mm) and dividing by the measured time, you will be able to determine how fast the action potential is being conducted down the nerve.

$$\text{Speed} = \frac{\text{distance (meters)}}{\text{time (seconds)}} = \frac{0.005}{\text{delta T}}$$

$$\textbf{Example:} \quad \frac{0.005}{0.00140} = 35.71 \text{ m/sec}$$

Experiment 3: Determining the Absolute and Relative Refractory Periods

In a single neuron, the absolute refractory period is the time following an action potential, during which the neuron membrane will not respond to any stimulus. This lack of response occurs because Na+ channels are (1) already open or (2) in the inactive state, meaning that the inactivation gate on the Na+ channel is closed preventing it from opening. Following the absolute refractory period is the relative refractory period. During this time, the neuron axon will produce an action potential if a stronger stimulus voltage is applied because some Na+ channels have reset (some are still inactive) and K+ channels are open, hyperpolarizing the cell.

The absolute refractory period in a compound nerve is the time period during which the second of two stimuli produces no visible 2nd CAP. The nerve is out of the absolute refractory period when the second stimulus produces a small but visible 2nd CAP. The relative refractory period, in contrast, is the time period that extends from the end of the absolute period until all axons in the nerve are again responsive to stimulation (i.e., all axons are completely out of their refractory periods and have responded to the stimulus).

During this experiment, two separate stimuli of the same strength (maximal voltage) will be applied to the nerve. A pair of stimulus artifacts will appear on channels 1 and 2. The second stimulus is the same stimulus strength as the first but it is applied to the nerve membrane after a short delay. The first stimulus will create a CAP (following the first stimulus artifact) while the second stimulus will be used to determine when the membrane is again responsive. If the second stimulus is applied when all the nerve axons are still in their absolute refractory periods, no 2nd CAP will be visible following the second stimulus artifact (A). However, if the second stimulus is applied at the very beginning of the relative refractory period, a few of the nerve axons will have come out of their absolute refractory period and respond, producing a small 2nd CAP (B). The first appearance of this 2nd CAP denotes the end of the absolute refractory period and beginning of the relative refractory period. Eventually the second stimulus will be applied after all the nerves are out of refraction (end of relative refractory period). At this point, the 2nd CAP will be maximal, and no further increases in amplitude will occur, because no additional nerves remain to add in (C). This marks the end of the relative refractory period.

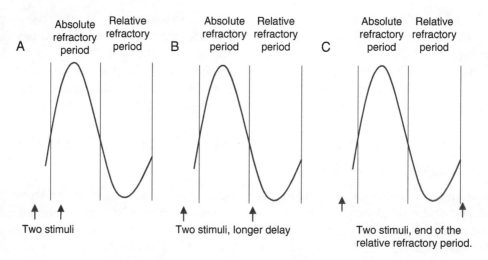

Setup

1. One member of your group should perform each of the following functions:
   a. **Director:** Reads the directions.
   b. **Controller:** Runs the computer.
   c. **Recorder:** Records data on the **Data Page** at the end of this section.
2. Log in to the computer at your station.
3. Turn the MP36 on using the switch at the back of the unit.
4. Double click on the BIOPAC BSL Student Lab 4.1 icon on the desktop.
5. Click on the PRO Lessons tab, select Action Potential and then click OK.
6. The screen shown in **Figure 3.1** should appear. Two channels, channel 1, which records the voltage of the stimulus applied to the nerve, and channel 2 that records the nerve response to that stimulus, should be apparent.

Calibration

1. Check to see that the stimulator is set to 1.0 volt at the bottom of your graph window (circled in **Figure 3.1**).
2. Click on Start. A stimulus signal should appear on channel 1 and a stimulus artifact and a CAP peak should appear on channel 2 (See **Figure 3.1**).
   • If no signals appear on either channel or if your CAP is very small, contact your instructor.
3. Save your data (File>Save as) in your section's folder within the Student Data folder on the desktop. Name your file Action Potential.

Experiment 1: Determining Threshold and Maximal Voltage

Procedure

I. Finding the Threshold Voltage

1. Reset your stimulator to 0.06 volts by highlighting the 1.0 and typing 0.06 in the volts box of the stimulator window at the bottom of the graph window. Hit Enter.
2. Click the Start button. A small stimulus signal should appear on channel 1 and a stimulus artifact (voltage change that results from the electrical signal traveling in the liquid outside the nerve to the measuring electrodes) on channel 2, but no CAP should appear on channel 2 (**Figure 3.2**).
3. Increase the voltage by 0.02, hit enter and then click on Start.

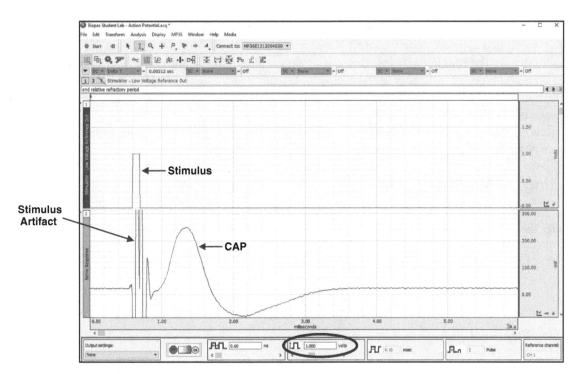

Figure 3.1

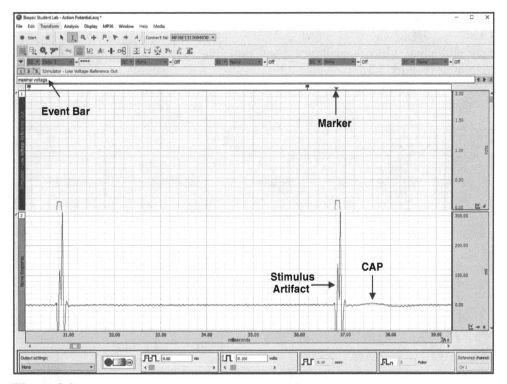

Figure 3.2

4. Repeat increasing the voltage in increments of 0.02 until a small CAP appears on channel 2 (**Figure 3.2**).
    - **If your computer fails to stimulate (no stimulus on channel 1 after you click *Start*), just stimulate again at the same voltage.**
    - **If you are not sure it is a CAP, apply another voltage increase and see if the CAP becomes larger.**

5. Insert a marker event above the stimulus by right clicking in the bar below the event bar. In the drop-down menu that appears, click on Insert New Event.
6. Right click **on the marker triangle** and click Edit event. Highlight and delete any writing in the marker text box, and type: Threshold (**Figure 3.2**).
7. Note the voltage setting on your graph window in the stimulator section. Record this as the threshold voltage on the data sheet at the end of this exercise.
   a. Record the time in milliseconds (at the bottom of the screen, just before your threshold stimulus and response). _____43 msec_____ milliseconds (ms)

II. Finding the Maximal Voltage

1. Increase the stimulus voltage by 0.10 volts above your threshold value. You may click the right arrow under the voltage window to increase by 0.10 volts at a time (**Figure 3.3**).

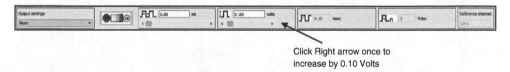

Click Right arrow once to
increase by 0.10 Volts

Figure 3.3

2. Click on the Start button. A larger amplitude CAP should appear on Channel 2 following a larger stimulus artifact.
3. Continue to increase your stimulus voltage by 0.10 volts and click on Start (once at each setting) until the CAP amplitude no longer increases significantly compared to the previous one. **DO NOT EXCEED 2.0 VOLTS!!**
4. Perform one additional 0.10 voltage increase and click Start. You should have a total of three CAPs of similar amplitude (**Figure 3.3**).
5. The last voltage setting that significantly increased the CAP amplitude over the previous voltage is the maximal voltage. Use the scroll bar arrow at the bottom of the screen to scroll backward to find your

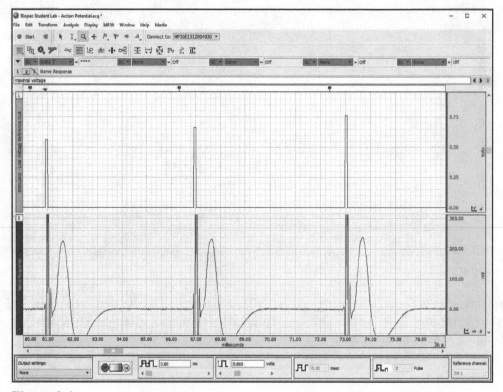

Figure 3.4

maximal CAP. Right click above the stimulus (in the bar below the event bar) to insert a new marker. When you do, your maximal voltage should be visible in the event bar. Record this value on the data sheet at the end of this exercise.

III. Viewing the Entire Voltage Versus CAP-Response

1. Scroll to the end of your data using the right arrow of the scroll bar at the bottom of the screen. This will set the upper time value to the end of your current data.
2. The lower time number must be manually entered to correspond with the time value at which you found threshold. To do so:
   a. Click on the milliseconds scale at the bottom of the graph window.
   b. A horizontal axis scale window should appear. Highlight the number in the box marked start and type in the milliseconds value that you recorded at threshold voltage. This should show the data from your threshold to the end of the maximal section.
   c. Click OK.
3. Print your data to the printer labeled **TS380 - HP1505N on MAVPRINT**. Make sure that the orientation is Landscape and the Fit Sheet on One Page box is selected.
4. On your printout, mark with an arrow on your graph your maximal CAP. Ensure that you are looking at the CAP amplitude, and not the stimulus artifact amplitude.
5. Save your data (File>Save).
6. Turn in your printout (It should look like **Figure 3.5** - be sure to put all group members' names on the graph) along with the data sheet at the end of this exercise to your instructor at the end of your lab session.

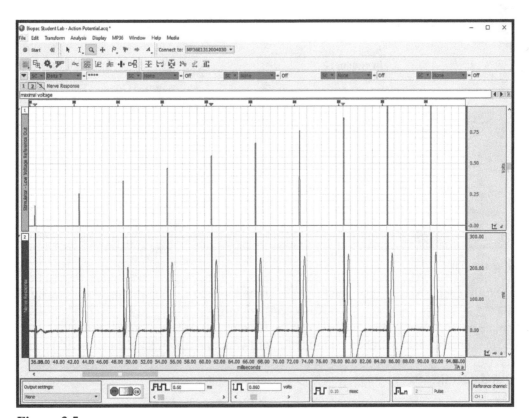

Figure 3.5

Experiment 2: Determining the Speed of Conduction of the Action Potential Down the Nerve

Procedure

1. Adjust your time scale to span 8 milliseconds. To do so, click on the milliseconds scale at the bottom of the graph window. Add 8 to the Start value and record this in the end box. For example, if you start value is 30 milliseconds, the value that you should record in the end box is 38 (30 + 8). Click OK.
2. Scroll to the end of your data using the scroll bar on the bottom of the graph window.
3. Make channel 3 visible by holding Alt and clicking on the 3 with the line through it (located on the top left of the graph window above the event bar). When you do, your last stimulus should be visible on channel 1 and your last CAP should be visible on channels 2 and 3.
4. To determine how fast the depolarization traveled from the first set of electrodes (channel 2) to the second set of electrodes (channel 3), click on the I-icon and drag to highlight a section from the peak of the last CAP in channel 2 to the peak of the CAP in channel 3 (**Figure 3.6**).

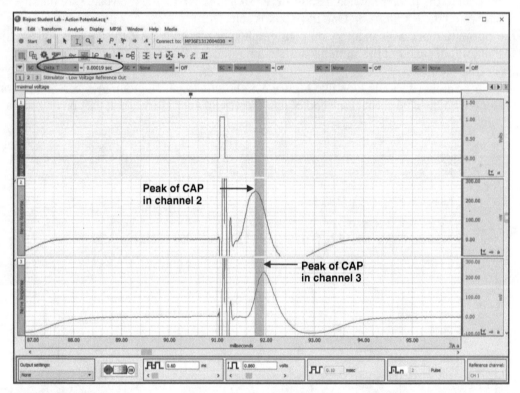

Figure 3.6

5. The time difference is shown in the box delta T, at the top left of the graph window (circled, **Figure 3.6**).
   Delta T = ___.00015___   Milliseconds = ___.15___
6. To calculate speed, take the distance between the channels (0.005 meters) and divide it by the delta T value (in seconds) that you recorded in #5. Record your answer on the data sheet at the end of this exercise.
7. Save your data (File>Save).

Experiment 3: Determining the Absolute and Relative Refractory Periods

Procedure

I. Determining the Absolute Refractory Period

1. Make channel 3 disappear by holding Alt and clicking on the 3 on the top left of the graph window above the event bar.
2. Lengthen the computer recording period (acquisition time) by clicking on the MP36 menu and then select Setup Acquisition.
3. Click on the Length/Rate option on the left hand side of the Setup window that appears (**Figure 3.7**).
4. Highlight the 6.020 in the Acquisition Length and type 15. **DO NOT CLICK THE RESET BUTTON!!**
5. Press Close.

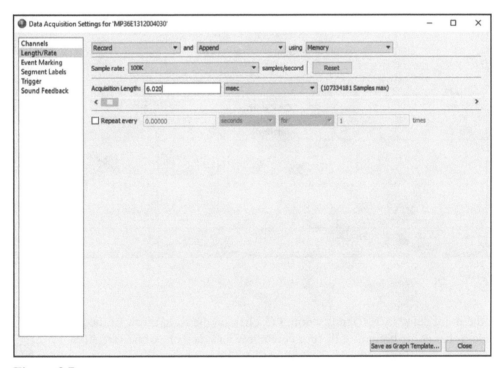

Figure 3.7

6. Adjust your time scale to span 30 milliseconds. To do so:
   a. Click on the milliseconds scale at the bottom of the graph window.
   b. A horizontal axis scale window should appear. Highlight the number in the box marked end and type in the start value + 30. For example, if your start value is 60 ms, then you would type 90 in the end value box.
   c. Click OK.
7. In the Output settings box on the lower left part of the graph window (in the stimulator control bar), click on the None and select Refr P in the drop down menu (circled in **Figure 3.8**).

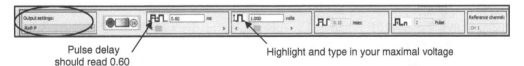

Pulse delay should read 0.60

Highlight and type in your maximal voltage

Figure 3.8

8. In the volts section of the stimulator bar, highlight the 1.000 and type in your maximal voltage (**Figure 3.8**).

9. In the pulse delay window, the stimulator bar should read 0.60 milliseconds (**Figure 3.9**).

   Note: The pulse delay sets the time delay between the first and second stimulus. A larger number means that more time lapses between the first and second stimuli.

10. Click on Start, and look for two stimuli on channel 1. You probably will only see one CAP on channel 2, and the second stimulus artifact will likely be on top the CAP (**Figure 3.9**).

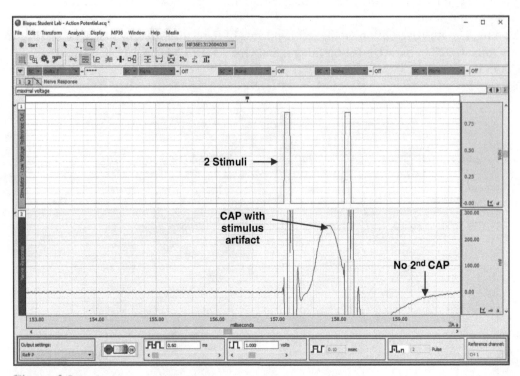

Figure 3.9

11. Increase the pulse delay by 0.10 milliseconds (1 click on the right arrow of the pulse delay scroll bar) in the stimulator bar at the bottom of the graph window and then click on Start. Look for a 2nd (but small) CAP on channel 2, following the 1st CAP. If no 2nd CAP is present, continue to increase the pulse delay and click on Start, until a second CAP is visible on channel 2 (**Figure 3.10**).

12. Insert a new marker event above the stimulus by right clicking in the bar below the event bar. In the dropdown menu that appears, click on Insert New Event.

13. Right click on the marker triangle and click Edit event. Highlight and delete any writing in the marker text box, and type: End absolute refractory. **DO NOT RECORD ANY VALUES ON YOUR DATA SHEET AT THIS TIME!!**

II. Determining the Relative Refractory Period

1. Increase the pulse delay by 0.3 milliseconds (3 clicks on the right arrow of the pulse delay scroll bar, in the stimulator bar at the bottom of the graph window) and click on Start. The 2nd CAP should have increased in amplitude.

2. Repeat step 1 until the 2nd CAP is about the same size as the *previous* 2nd CAP (**Figure 3.11**).

3. Insert a new marker event above the stimulus by right clicking in the bar below the event bar. In the dropdown menu that appears, click on Insert New Event.

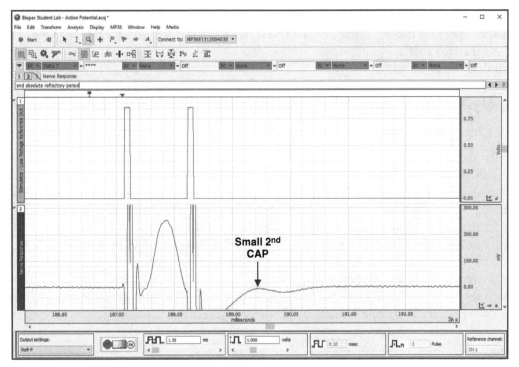

Figure 3.10

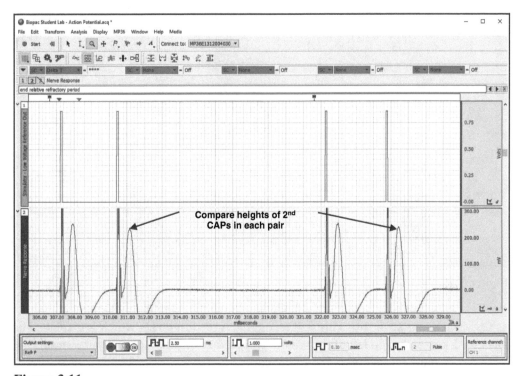

Figure 3.11

4. Right click on the marker triangle and click Edit event. Highlight and delete any writing in the marker text box, and type: End relative refractory. **DO NOT RECORD ANY VALUES ON YOUR DATA SHEET AT THIS TIME!!**
5. Save your data (File>Save).

III. Data Analysis – Measuring the Absolute and Relative Refractory Periods

A. Measure the absolute refractory period

1. Click on the second icon from the left under the Start menu and then on journal in the drop down menu that appears (See Exercise 1, Figure 1.5).
2. In the journal window that appears at the bottom of the graph window, type Absolute Refractory Period and hit Enter.
3. Adjust your time scale to span 10 milliseconds (See Experiment 2, step 4a – c).
4. Find your absolute refractory period marker using the left and right arrows (◄◄►) on the right hand of the screen.
5. Measure the absolute refractory period. To do so:
   a. Click on the I-icon and drag to highlight a section on channel 1 from the very beginning of the first stimulus to the very beginning of the second stimulus (**Figure 3.12**).

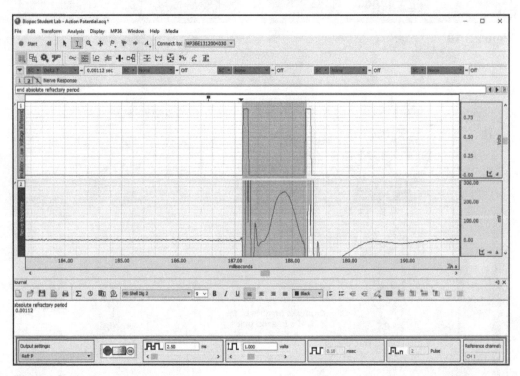

Figure 3.12

   b. Press Ctrl-M on your keyboard. This will record the time difference, which is displayed in seconds in the delta T box at the top left of the window (circled in **Figure 3.6**), in your journal. This value is equal to the absolute refractory period of the nerve.
   c. Record this value on the data sheet at the end of this exercise.

B. Measure the relative refractory period

1. Click in the journal, type Relative Refractory Period and hit Enter.
2. Find your end relative refractory period marker using the left and right arrows (◄◄►) on the right hand of the screen.

3. Mark the end of the absolute refractory period on the graph as follows:
   a. Click at the beginning of the first stimulus and drag until the value in the delta T box at the top left of the window is equal to your previously measured absolute refractory period (**Figure 3.13**).
   b. Insert a new marker event at the end of the absolute refractory (which is the beginning of the relative refractory period) by right clicking in the bar below the event bar.
4. To measure the relative refractory period, drag to highlight a section on channel 1 from the marker you just inserted, to the beginning of the second stimulus of that set of two stimuli (**Figure 3.14**) using the I-icon.

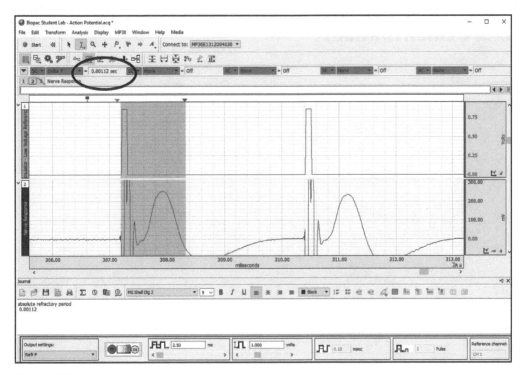

Figure 3.13

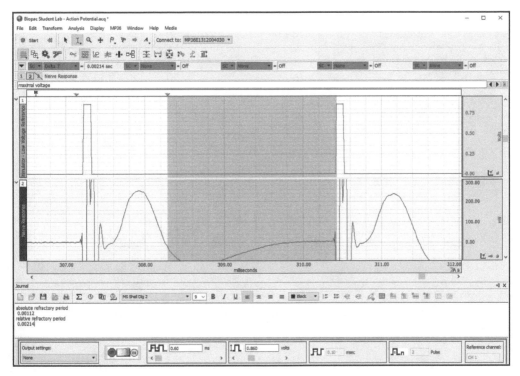

Figure 3.14

5. Press Ctrl-M on your keyboard. This will record the time difference, which is displayed in seconds in the delta T box at the top left of the window (circled in **Figure 3.13**), in your journal. This value is equal to the relative refractory period of the nerve.
6. Record this value on the data sheet at the end of this exercise.
7. Save your data (File>Save).
8. Close the MP36 graph window and turn the Biopac off via the switch in back of the MP36 unit.

Experiment 4: Effect of a Na$^+$ Channel Blocker on the Nerve (Instructor Demonstration)

Action Potential Data Sheet

Group Number: _____   Name: _____

Experiment 1: Threshold and Maximum

Threshold voltage: _____

Maximal voltage: _____

Why does the CAP get bigger with increased voltage stimulation? How does this compare to an action potential in a single axon?

Experiment 2: Speed

Speed of conduction: _____

Predict and explain what would happen to the speed of conduction if you were measuring unmyelinated axons.

Experiment 3: Refractory Period

Absolute refractory period: _____

Relative refractory period: _____

What causes the absolute refractory period? Can a second action potential be generated during this time?

What causes the relative refractory period? Can a second action potential be generated during this time?

Experiment 4: Lidocaine

What happened to the CAP after the lidocaine was added to the nerve?

What happened after the lidocaine was rinsed away?

Special Senses

Study Questions

1. What are modalities? Give some examples of receptors and their modalities.
2. How does the brain tell a cold stimulus from a warm one?
3. What is temperature adaptation (resetting of the zero)? How did we demonstrate this in lab?
4. How is the density of touch receptors determined? How is this related to the size of a receptor field, and the ability to localize a stimulus?
5. What is the relationship between density of touch receptors and the sensory and motor cortexes?
6. Which area had the greatest density of touch receptors? Which area had the lowest density?
7. Give one example of referred pain and explain why this phenomenon occurs?
8. Which receptors are involved in vision?
9. Which parts of the eye refract light? Which can adjust its refractory power?
10. What is accommodation? Why is it important in vision?
11. How does the lens of the eye alter its refractive strength?
12. Define each of the following:
    a. Myopia.
    b. Hyperopia.
    c. Emmetropia.
13. Which visual disorder is described to have an eyeball that is too long? Which is described to have an eyeball that is too short?
14. How did we test visual acuity? What chart did we use?
15. What does a visual acuity of 20/40 mean?
16. What is astigmatism? How did we test for it?
17. How did we measure the near point of vision?
18. As they age, why do people develop a near point of vision that is farther from their eye? What is the name for this condition?
19. Explain why we have a blind spot.
20. Define retinal disparity. How does retinal disparity help vision?
21. How did you test for nystagmus? What may cause this to occur in a patient as they watch a moving object?
22. Explain the process of phototransduction.
23. What are rods and cones? Which is used for color vision?
24. Why are rods more sensitive to low levels of light than cones?
25. Does a rod or cone have greater visual acuity? Why?
26. What causes positive and negative afterimages? How did we see these in lab?
27. What colors do the three different types of cones respond to?
28. What happens when a person lacks one of the cone types?
29. Explain why staring at a red square results in a negative afterimage of blue-green.
30. What receptors are involved in hearing?

31. Trace all the steps between the arrival of sound waves at the external ear to the generation of an action potential in the cochlear nerve.
32. How is pitch determined by the brain? How is loudness determined?
33. What causes conduction deafness? What causes sensory deafness?
34. How does a hearing aid work? Can it correct for both types of deafness?
35. Why was the sound heard louder in the plugged ear when you were using the sound byte?
36. What type of receptor is used to help you maintain a sense of balance and equilibrium?
37. Why is vestibular nystagmus initiated when a rotating subject is suddenly stopped?
38. How does information in the vestibular system become an action potential in the vestibular nerve?

Applied Questions (Answers in Back)

1. Tory Tense has muscle spasms in her left shoulder. She also has tingling in her left hand, yet her left hand appears normal. Is the problem in her hand, or in her shoulder? Explain.
2. Problems with the reproductive organs in women often lead to complaints of lower back pain. Explain in terms of referred pain why this may be a common complaint.
3. Peter has spilled acid on his hand. He said it burns, yet the acid was at room temperature. Explain in terms of stimulating a receptor with an alternate modality, how this sensation occurred.
4. Jane's grandmother tends to hold papers very far away as she attempts to read them. She most likely has what visual disorder?
5. A person who focuses light in front of the retina would have which visual disorder?
6. Which cones would be stimulated by the color purple? By the color aqua?
7. Jill's closet light is burned out. The closet is only dimly lit from the room light. When Jill arrived at school, she realized she was wearing one blue shoe and one black shoe. Why did this happen?
8. Roger's grandfather has had a Weber's hearing test performed because he can't hear well out of his left ear. The tuning fork sound was louder in his left than in his right ear. What type of deafness does he have?
9. Action potentials on the cochlear nerve are arriving at the brain with a very high frequency, and are arriving primarily from nerves in the cochlea far from the oval window. What type of sound is this?
10. Uncle Joe has been diagnosed with an inner ear infection. He feels dizzy and nauseous and asks you to explain why he feels like the room is spinning. (Hint: action potentials are generated in the vestibular nerve–how are these interpreted by the brain?)
11. During a high fever, patients complain of a "ringing in the ears." Explain this phenomenon in terms of the physiology of the sensory system and the brain's interpretation of action potentials on the cochlear nerve.

Introduction

Sensory

The **sensory system** helps the CNS process information about the internal and external environment, and includes the senses of: touch, vision, hearing, taste, and smell. Indeed, without the sensory system, we would not be able to perceive anything about our environment. The brain receives information from sensory neurons in the form of action potentials. These action potentials are interpreted by the brain using: 1) **type** of receptor stimulated, 2) the area of the body stimulated (**location coding**), 3) the frequency in which the action potentials are being delivered (**frequency coding**), and 4) the thresholds of all receptors stimulated (**population coding**).

The sensory system is composed of specialized receptors that are stimulated by only certain types of energy from the environment. For instance, the receptor in the eye is stimulated by light energy, but cannot be stimulated by sound waves. The type of stimuli that stimulates a specific receptor is called the **modality**.

Examples of modalities include: sound, light, chemicals, and temperature. Receptors produce graded potentials, which can create action potentials in attached sensory neurons if the neuron is brought to threshold. The action potentials are eventually carried to specific regions of cortex, depending upon the type of sensory receptor stimulated. For example, tactile information is routed specifically to regions of the postcental gyrus, while visual information is routed to cortical regions in the occipital lobe. Specific areas of the cortex, then, are devoted to the perception of specific sensations. Indeed, if an individual's visual cortex (occipital lobe) was artificially stimulated with an electric probe, the individual would perceive ("see") light, even though there was no light present.

Somatosensory System

The somatosensory system is involved in interpreting various sensations in the skin (Ex. touch, pain, temperature). For example, certain receptors in the skin detect changes in skin temperature. These receptors are scattered throughout the skin, and some thermal receptors respond to higher temperatures while others respond to lower temperatures. It is important to note, that thermal receptors detect <u>changes</u> in skin temperature. Thus, when a person first enters a hot tub skin thermal receptors detect an increased temperature, and fire a high frequency of action potentials to the brain (which registers the sensation as hot). However, after a few minutes, the thermal receptors "adapt" to this higher temperature by propagating fewer action potentials, and the brain stops perceiving the sensation of heat. If a person then leaves a hot tub and jumps into a pool, the thermal receptors must then readapt to an even lower temperature (and the pool water "feels" quite cold relative to the warm hot tub). However, once the thermal receptors have adapted to the cooler water temperature, the water will no longer feel as cold. Also note, that extremely high and low temperatures will also stimulate pain receptors (nociceptors) resulting in a burning sensation.

The sensation of "touch" is another aspect of the somatosensory system. Mechanoreceptors, which respond to pressure, are involved in sensing touch. The closer touch receptors are to one another, the greater their density, and the more sensitive that area of skin is to touch. Cells expressing touch receptors have long axons that enter the spinal cord and eventually make synapses in the brain. A **sensory homunculus** is a map of the primary somatosensory cortex (postcentral gyrus) showing areas of the cortex devoted to sensations from various regions of the body. The size of the area in the sensory cortex devoted to a specific body location is related to the density of receptors in the body area. Body areas (such as the lips) that have a high density of receptors have a large dedicated area in the sensory cortex. Body areas (such as the legs/trunk) that have a low density of receptors have a smaller dedicated area in the sensory cortex.

The somatosensory system is also involved in sensing pain, which can sometimes be a complex experience. Sometimes an individual may "feel" pain in one area of the body but the actual source of the pain is located elsewhere. This phenomenon is known as **referred pain**. Referred pain is common when the internal environment or the viscera (internal organs) is involved. Pain information from nociceptors in the viscera is often *referred* to the body surface. Sensory neurons from the viscera and sensory neurons from the body surface enter the spinal cord together, and synapse onto nearby secondary neurons. Typically, the body's surface sends more signals to the brain than does the viscera. Thus, as the brain is accustomed to receiving sensations originating from the body surface, it therefore misinterprets the visceral information as coming from the body surface. This is why a person who has a heart attack may feel heaviness in the left side of the chest or pain in the left arm and shoulder. Physicians often use symptoms from referred pain to locate which organ may be troubling the patient. Another type of referred pain occurs when a nerve is stimulated somewhere along its pathway to the brain. The resulting action potential is misinterpreted by the brain as coming from the origin of the nerve. Thus, patients that rupture a spinal disk (in the back) that stimulates sensory nerves from the leg *feel* like their leg aches.

Visual System

The eye contains photoreceptors that take light energy (photons) and convert it into electrical signals that are sent to the visual cortex. Photoreceptors are located in the retina of the eye. For clear vision to occur, light rays as they enter the eye must bend and come together at a single point on the retina. The bending of light rays is known as **refraction**. Three structures in the eye refract light: **cornea**, **vitreous**, and **lens**. The cornea bends light rays the most, so it is said to have greatest refractive power. However, the lens can not only bend but <u>adjust</u> the bending of the light rays to assure that the rays converge onto the retina. The thicker the lens becomes, the more it will bend the light rays. This ability of the lens to thicken and focus on nearby objects is called **accommodation**.

When viewing an object close up, less distance exists between the object and the retina, and the rays diverge as they enter the eye. Therefore, the lens must thicken significantly to bend these rays sharply so that they converge on the retina. On the other hand, for distant objects, there is a greater distance between the object and the retina, so the light rays enter the eye relatively parallel. In this case, the lens will flatten (thin) - appropriately decreasing its refractive power. A person with normal vision (**emmetropia**) has an eye lens that can accommodate for different incoming light ray angles so that they appropriately converge onto the retina (where the photoreceptors are located). Accommodation can be experienced by either focusing on a screen held near your eye or focusing on a distant object through the screen. Note that the lens can only "focus" on either a near object **OR** a far object- not both. Therefore, either the distant object or the screen will be out of focus.

Visual defects commonly occur when light rays do not appropriately converge/focus onto the retina, and therefore vision is "blurry." In **myopia** (near sightedness), light rays converge in front of the retin, and the individual will not clearly see objects that are far away. Myopia can be caused by an eyeball that is too long, and thus the lens is "too strong" for the eye. To treat myopia, a concave lens is used to prevent the light rays from bending too sharply.

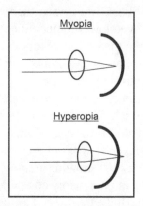

In a different condition, **hyperopia** (far sightedness), light rays converge behind the retina when the person attempts to see a nearby object. This is sometimes described as an eyeball that is "too short." Since the refractive power of the lens is too weak in this case, a convex lens will be needed to bend the rays more sharply so that they converge appropriately onto the retina. Note that since light rays from a distant object come into the eye in a relatively parallel fashion (and they don't need to be bent so much), a person with hyperopia can see distant objects just fine.

The ability of the eye to appropriately focus light onto the retina is called **visual acuity**. Visual acuity is typically tested using a **Snellen eye chart**. The top number in a Snellen eye chart indicates how far from the chart your partner is standing, while the bottom number represents how many feet from the chart a person with normal vision could stand yet still read the same lowest line as your partner. For instance, a vision of 20/20 means your partner can read the line at the same distance as a person with normal vision (a visual acuity of 20/20 means your partner has normal vision). A visual acuity of 20/15 means your partner can read a line 20 feet away that a normal person can only see within 15 feet of the line; thus your partner would have better than normal vision.

An **astigmatism** occurs when either the cornea or lens has an abnormal curvature or other irregularity, causing part of an object to be in focus while another part is out of focus.

Presbyopia or "old eyes" is caused by stiffening of the lens as people age. Since the lens is more stiff (less pliable) it can no longer thicken/bulge enough to focus on nearby objects, and the "near point of vision" moves outward (the **near point of vision** refers to how close an object can be brought to the eye yet still be in focus). Thus, for a person with **presbyopia**, the near point of vision is longer (farther away) than for someone with normal vision. That is why as people age, and develop presbyopia, they tend to hold newspapers and books further away from their eyes- so they can read them. This can be corrected using "reader glasses."

The **pupil** helps regulate the amount of light entering the eye, and therefore protects the photoreceptors from becoming over stimulated. To adjust for the amount of light hitting the retina, pupils constrict in bright light, and dilate in dim light. This **pupillary reflex** is performed via the autonomic nervous system (ANS). Activation of the parasympathetic branch of the ANS tends to "shrink" the pupil; while activation of the sympathetic branch (parasympathetic off) tends to increase pupillary size (think- wide eyes when frightened). Interestingly, the parasympathetic nervous system also innervates muscles that control lens thickening. Thus, when looking at a nearby object, the parasympathetic nervous system is activated to thicken the lens to appropriately accommodate the light and focus it on the retina. Concomitantly, however (as the parasympathetic is activated) focusing on a nearby object also causes the pupils to shrink.

When photoreceptors in the retina become activated, they send action potentials via the optic nerve towards the visual cortex. The optic disk is the area of the retina where the optic nerve and blood vessels exit the eye. Interestingly, the optic disk contains no photoreceptors. Thus, if light enters the eye and happens to be focused on the optic disk, no action potentials will be generated (and the image will not be seen). In other words, there is a **blind spot** on each retina - located at the optic disk.

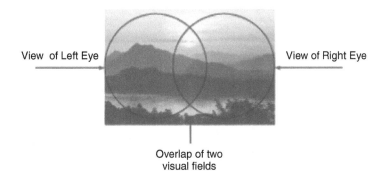

View of Left Eye   View of Right Eye

Overlap of two
visual fields

Humans have **binocular vision**, which means we see one unified image with two eyes. Your brain takes information from both eyes and overlaps the middle portion of each visual field while maintaining the lateral visual field of each eye. **Retinal disparity** occurs at the overlap of the two visual fields- as the eyes are laterally displaced from each other by a few centimeters. Retinal disparity is responsible for three-dimensional vision and depth perception, since each eye sees an object from a slightly different angle. Indeed, an individual with vision in only one eye would have a difficult time walking down an unfamiliar flight of stairs because he/she would have poor depth perception.

In order to see a unified picture of the world, your eyes must be positioned so that they can work together in a coordinated fashion. Correct positioning of the eyes depends on strong ocular muscles. If ocular muscles are weak, the eyes may twitch or jerk when they move; this is known as **nystagmus**. Weak eye muscles may prevent some individuals' eyes from working in a coordinated manner when they read, and people with nystagmus report that words appear as if they are "falling off the page." Thankfully, nystagmus can often be corrected by exercising the ocular muscles.

Phototransduction is the process of converting light energy into electrical energy (action potential) that is sent to the brain for interpretation. Light is absorbed by a photopigments (Ex. Retinal) that are complexed to a "G-protein receptor" (Ex. Opsin) embedded within the plasma membrane of photoreceptor cells (rods and cones) of the retina. When light hits the photopigment, the pigment changes shape and disassociates from its G-protein receptor (opsin separated from retinal becomes "bleached opsin" which is no longer responsive to light). This causes the G-protein receptor to change conformation, and activates a G-protein signaling cascade which includes cGMP and Na^+ channel conformational changes. These events ultimately result in a change in membrane potential of the photoreceptor cell. Photoreceptor membrane potential changes eventually trigger action potentials in optic nerve axons that travel to the occipital lobe of the brain to be interpreted as "vision."

There are two main types of photoreceptors in the retina, rods and cones. **Rods** are responsible for black and white vision in dim light (night) situations. Notice in **Figure 4.1** that several rods synapse onto a single bipolar cell and several bipolar cells synapse onto a single ganglion cell (optic nerve). Because of this "convergence," input from several rods can add together to stimulate the bipolar cell, requiring less light energy for a signal to be sent to the brain. Thus, rods are <u>more sensitive</u> to low light levels and help us see in dim light situations (Ex. nighttime). However, as several rods send information via a single axon to the brain, vision is not as sharp (<u>lower acuity</u>).

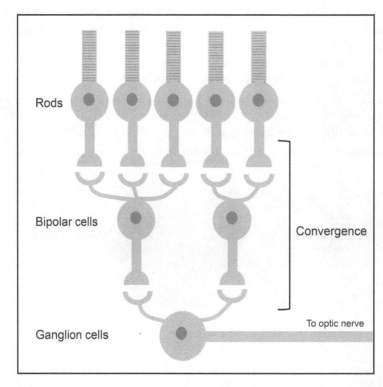

Figure 4.1

Cones are involved in color vision that occurs in a bright light situation (Ex. daytime). **Figure 4.2** illustrates that each cone synapses with only <u>one</u> bipolar cell, which synapses with only <u>one</u> ganglion cell. In other words, cones do <u>not</u> converge. Thus, each cone sends information through a single axon to the brain, and the vision is <u>sharper</u> (higher acuity). However, a single cone needs more light energy to adequately stimulate a bipolar cell, since stimuli (from neighboring cones) cannot be "added up" within the bipolar cell. Therefore, cones are <u>not as sensitive to light</u>, and do not function as well as rods in low light conditions. There are three main types of cones (<u>red, green, and blue</u>) based on the precise type of visual pigment each contains. According to the **Young–Helmholtz** theory of color vision, other colors in the visual spectrum result from the brains interpretation of the combined signaling of the different cone types. If one cone type is defective, the individual will be **color blind**, and have difficulty distinguishing certain colors. Color blindness is more common in males than females, since genes encoding cone photopigments are located on the X-chromosome.

If you look at a bright light, then turn away, you will still see a "halo" of the white light for a short time. This is called a **positive afterimage**, and it results from the photoreceptors continuing to depolarize and send information to the brain. After a short period of time, the light turns into a black spot. This is a **negative afterimage** which occurs because the photoreceptors with "bleached opsin" cannot respond to the new visual field until the photopigment is regenerated. When the photopigment of a single cone type becomes *bleached*, the afterimage will be a combination of the two cone types that were <u>not stimulated</u>, and therefore can still fully respond to light.

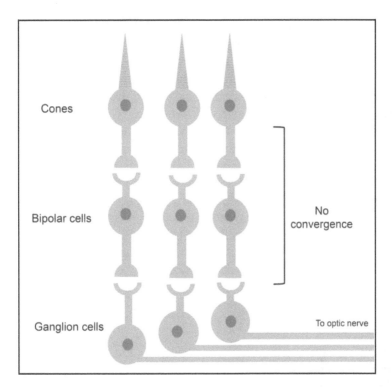

Figure 4.2

Auditory System

The process of converting sound waves into electrical signals that are then interpreted by the brain is called **auditory transduction**. Auditory transduction initiates in the **external ear** (pinna and ear canal) where sound waves are directed toward the tympanic membrane (ear drum). The **middle ear** consists of the tympanic membrane and a series of three attached bones called ossicles. These ossicles are also connected to both the tympanic membrane and to the oval window of a structure in the **inner ear** called the cochlea. The **cochlea** is the true site of auditory transduction, and is a fluid-filled structure that contains the **organ of corti**. The organ of corti consists of a basilar membrane on which hair cells sit, and a tectorial membrane directly above the hair cells. When a sound wave strikes the tympanic membrane, the tympanic membrane vibrates. This vibration causes the ossicles to push and pull on the oval window, creating waves in the fluid of the cochlea. These cochlear waves undulate the basilar membrane causing the hair cells to press and deflect up against the tectorial membrane. **Bending of the hair cells** opens mechanically gated cation channels (TRPA1 channels) located within the hair cell plasma membranes. **Cation (K^+) influx** causes hair cells to depolarize, which leads to Ca^{++} influx and consequent glutamate exocytosis from the hair cell onto the cochlear nerve. Excitation of the cochlear nerve ultimately leads to action potentials that travel towards the auditory cortex of the brain for "interpretation" of sound.

The brain interprets the **pitch** of a sound by the location of which hair cells are being stimulated. High pitch sounds will maximally stimulate hair cells located near the oval window. Low pitch sounds will maximally stimulate hair cells further from the oval window. The brain interprets the **loudness** of a sound by the frequency of action potentials generated by the hair cells. A soft sound will only make small amplitude waves in the cochlear fluid, and therefore bend hair cells just a little bit (only letting in small amounts of K^+). Modest amounts of hair cell bending will only allow modest amounts of glutamate exocytosis, which ultimately creates fewer action potentials being generated in the cochlear nerve per second (lower frequency of action potentials). A loud sound will create larger waves in the fluid filled cochlea (causing the hair cells to bend more), ultimately producing more action potentials per second (higher frequency of action potentials) being generated in the cochlear nerve.

Conduction deafness is the reduced ability to hear, due to a problem between the outside of the ear and the cochlea. Causes of conduction deafness include external ear problems or middle ear damage such as: build up of wax in the ear canal, ear infections, or damage to the ossicles. When the middle ear is damaged, waves may not be created in the fluid-filled cochlea by sound waves entering the ear canal. However, if the hair cells are still healthy, the brain is still capable of detecting sound. If the sound can be conducted through the bones in the skull, the vibrations can bypass the middle ear and still create waves in the cochlear fluid. Individuals with conduction deafness may improve their hearing by wearing a hearing aid–which amplifies incoming sounds and conducts sound waves through the bone to the inner ear, bypassing the external and middle portions of the ear. **Sensory deafness** (sensorineural deafness) occurs when transmission of nerve impulses from the cochlea to the brain is impaired. This is usually due to damage to the cochlea or the vestibulocochlear nerve. Prolonged exposure to loud sounds, infections, and some drug toxicities can lead to sensory deafness. A hearing aid will also aide people with sensory deafness, however some individuals may require a cochlear implant.

Vestibular System

Your vestibular system helps you maintain a sense of balance and equilibrium by using a set of inner ear structures called: **semicircular canals**, **utricle**, and **saccule**. This system helps detect linear and rotational acceleration, using hair cell receptors (similar to hearing). Like the cochlea, the semicircular canals are filled with fluid, and contain hair cells that have projections called **stereocilia**. When fluid in the semicircular canals is in motion, it bends the stereocilia and creates a change in the rate of action potentials generated in the attached nerves. This "frequency coded" information travels to the brain on the vestibular nerve, and the brain interprets the information as motion. When an individual spins around, the head starts to rotate with the rest of the body, but the fluid in the semicircular canals (endolymph) initially lags behind. The endolymph, therefore, bends the stereocilia (in a direction opposite the spin) changing the rate in which the attached afferent neurons generate action potentials. The brain then perceives spinning. Eye movements and other changes in body position will also occur to help the body compensate and maintain its orientation in 3D space. Once the spinning becomes stable (no acceleration), the endolymph no longer lags behind and the stereocilia become upright again. As a result, action potential frequency again slows. However, when a person stops spinning (head stops moving) the endolymph will continue moving in the direction of the spin for a while. This again will bend the stereocilia (in the opposite direction of the original spinning). The frequency of action potential generation will change once again, and the brain will perceive a spinning sensation (even though spinning has stopped). This will continue until the endolymph stops moving and stereocilia go back to their upright position.

Vertigo is the illusion of spinning when the body is not actually rotating. It is caused by activation of the vestibular system by means other than rotational acceleration. When the vestibular system is over stimulated, it can activate the autonomic nervous system that may cause nausea and vomiting. Inner ear infections can cause inflammation or a buildup of fluid in the semicircular canals, which bends the hair cells and causes vertigo. The movement of the eyes back and forth in an oscillatory manner (which can occur when traveling in a car or on a boat) can also over stimulate the vestibular and autonomic systems and cause **motion sickness**. Interestingly, the vestibular system is neurologically connected to the visual system so that the body can adjust and reorient itself when the vestibular system is signaling. **Vestibular nystagmus** is a phenomenon in which the eyes quickly move (jerk) in response to an over-activated vestibular system (Ex. spinning in a chair).

Procedures

I. Examining the Somatosensory System

Experiment 1: Mapping Thermoreceptors

1. Draw or imagine a 1-inch square on your partner's forearm, a few inches below the wrist.
2. Pick up a hot probe from the water bath in the back of the room and a cold probe from cooler next to the water bath. Dry both probes with a piece of paper towel.
3. Have your partner close his/her eyes while you touch the skin on the forearm with the hot or cold probe in various spots within the square. Make sure you alternate touching the skin with the hot and cold probes.
4. Mark an H where your partner feels hot and a C where your partner feels cold. Draw this map on your data sheet at the end of this exercise.

Experiment 2: Adaptation of Thermal Receptors

1. In the back of the lab are three containers of water. One container has cold water, one has lukewarm water, and the other hot water. Place your right hand in the hot water and your left hand in the cold water and keep them there for 1 minute.
2. Place both hands in the lukewarm water and tell your partner which hand feels the water as being warmer.
3. Record your observations on the data sheet at the end of this exercise.

Experiment 3: Two-point discrimination test

1. With the two points of adjustable calipers far apart, place both points on the skin of your partner's palm, touching at the same time with equal pressure on each point. Make sure your partner is not watching what you are doing.
2. Keep bringing the points closer together until your partner can only feel one point.
3. Measure and record the smallest distance between the two points in which your partner could feel both points, and record this measurement on the data sheet at the end of this exercise.
4. Repeat steps 1–3 on the 1) back of the hand, 2) the fingertip, and 3) the back of the neck.

Experiment 4: Referred pain

1. Gently tap the medial side of your partner's elbow with a mallet until you hit the ulnar nerve - your partner will notify you when you have hit the correct spot.
2. Record where he/she feels the sensation on your data page at the end of this exercise.
3. Based on your results, where do you think the ulnar nerve originates?

II. Examining the Visual System

Experiment 5: Accommodation

1. Hold a small screen in front of your eyes.
2. Focus on the screen and notice how objects through the screen appear blurry.
3. Now focus on a distant object through the screen, and notice how the screen appears blurry.
4. Record your observations on your data sheet at the end of this exercise.

Experiment 6: Visual acuity

1. Have your partner stand 20 feet away from the Snellen eye chart and cover one eye.
2. Your partner should read the smallest line that he/she can see. He/she can miss up to two letters on that line.
3. Record his/her visual acuity (*e.g.*, 20/20 - on the side of the smallest line that your partner was able to read) in the data sheet at the end of this exercise.

Experiment 7: Astigmatism

1. Have your partner stand 20 feet away from the astigmatism chart. The chart is a black circle with black lines going toward the center.
2. Your partner should cover one eye and determine if any lines are darker than others. If so, your partner has astigmatism.
3. Repeat the procedure for the other eye.
4. Record your results on the data sheet at the end of this exercise.

Experiment 8: Near point of vision

1. Hold a ruler under one eye so it is facing outward.
2. Gradually bring a pencil toward your eye until the eye can no longer focus on it.
3. On your data sheet at the end of this exercise, record the distance in centimeters from the object to your eye.
4. Repeat the procedure with your other eye.

Experiment 9: Pupillary reflex

1. Shine a flashlight to the side of your partner's face. While observing the pupil, move the light so it shines into the eye.
2. Next, hold a pencil several feet away from your partner and slowly bring it toward his/her face.
3. Note what happens to pupil size as you bring the pencil closer to his/her eye.
4. Record your observations from both experiments on your data sheet at the end of this exercise.

Experiment 10: Blind spot

1. Hold the figure above slightly below and in front of your face, arm's length away from your eyes. The dot should be in front of your right eye and the cross should be in front of your left eye.
2. Close your right eye and focus on the dot. Slowly bring the figure closer to you until the cross disappears from your peripheral vision. NOTE: The cross will disappear then reappear again quickly, so make sure to bring the figure slowly toward your eyes or you will miss it.
3. When the cross disappears, stop moving the figure and record the distance between the figure and your eyes. Record the distance on the data sheet at the end of this exercise.
4. Repeat the procedure with the left eye closed, but turn the book upside down. The cross should now be to your right.

Experiment 11: Retinal Disparity

1. Roll a piece of paper into a tube and place it over your right eye.
2. With both eyes looking straight forward, take your left hand and run it in front of the tube.
3. What do you see? Record your observations on your data sheet at the end of this exercise.

Experiment 12: Visual Nystagmus

1. Take a pencil and have your partner follow it with his/her eyes (without moving his/her head).
2. Slowly move the pencil and make a large Z. Observe your partner's eyes for any twitching.
3. Make a large X with the pencil, and a large circle. Bring the pencil slowly from farther away to the front of your partner's face, until it touches his/her nose. Watch your partner's eyes for twitching or jerky movements.
4. On the data sheet at the end of this exercise, record whether or not you observed nystagmus.

Experiment 13: Color blindness

1. Test for color blindness using Ishihara's test booklet in the back of the lab.

Experiment 14: Afterimage

1. Look at a certain colored square for 30sec (to bleach a certain cone type).
2. Look at a white square (all colors of light)
3. Record what type of afterimage (glow) that you see in your data sheet at the end of this exercise.

III. Examining the Auditory System

Experiment 15: Rinne's test

1. While holding the handle with two fingers and the thumb, strike a tuning fork on the lab bench. Place the handle of the tuning fork against the mastoid process behind the ear with the fork pointed downward (**Figure 4.3**). When the sound has just faded, move the tuning fork near the opening of the ear canal (**Figure 4.4**). If the sound "reappears," the person has no conduction deafness. Record your results in your data sheet at the end of this exercise.

Figure 4.3

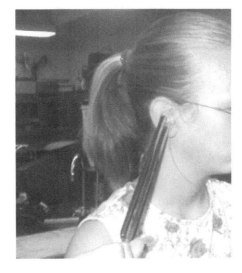

Figure 4.4

Experiment 16: Weber's test

1. While holding the handle as before, strike a tuning fork on the lab bench. Place the handle on the mid-sagittal line of the skull (**Figure 4.5**) and note whether you hear the sound equally in both ears.

2. Repeat the procedure, but plug one ear. In which ear is the sound louder? When you plug one ear, or have conduction deafness, background noises from the environment are muffled, and the sound appears louder in the affected ear. Record your observations on the data sheet at the end of this exercise.

Figure 4.5

Experiment 17: Sound bytes (These transmit sound waves through your teeth and skull)

1. Pick up a sound byte and press one of the buttons in the front of the box. Do you hear any sound?
2. Take a sucker and put the sucker stick in the top of a sound byte.
3. Clamp down gently on the sucker with your teeth. Press the same button on the sound byte box. Do you hear any sound?
4. Plug one ear to mimic conduction deafness. Repeat steps 2 and 3. What happens to the sound in the plugged ear? Why?

IV. Examining the Vestibular System

Experiment 18: Vestibular Nystagmus

1. Have the student sit in a swivel chair with feet slightly off the floor.
2. The student's head should be down (chin to chest), and eyes should be closed.
3. Rotate the chair in one direction quickly about 10 times.
4. Suddenly stop the chair and have the student open their eyes wide so the nystagmus can be observed by classmates.
5. Record your observations on your data sheet at the end of this exercise.

Special Senses Data Sheet

Group Number: _____   Name: _____

I. Examining the Somatosensory System

Experiment 1: Mapping Thermoreceptors

Mark hot locations in one square and cold in the other

H = Hot   **C = Cold**

Experiment 2: Adaptation of Thermal Receptors - What did you experience?

Experiment 3: Two-point discrimination (Distance at which only one touch was felt.)

Palm _____ mm   Back of hand _____ mm

Fingertip _____ mm   Back of neck _____ mm

Experiment 4: Referred pain - What did you feel?

II. Examining the Visual System

Experiment 5: Accommodation – Describe your experience

Experiments 6, & 7: Visual acuity & Astigmatism

		Student 1	Student 2	Student 3
Visual acuity	Right eye	20/	20/	20/
	Left eye	20/	20/	20/
Astigmatism (Yes/No)	Right eye			
	Left eye			

Experiment 8: Near point of vision

		Student 1	Student 2	Student 3
Near Point of Vision	Right eye	cm	cm	cm
	Left eye	cm	cm	cm

Experiment 9: Pupillary Reflex: - What happened?

Experiment 10: Blind Spot

		Student 1	Student 2	Student 3
Blind spot (distance from face)	Right eye	cm	cm	cm
	Left eye	cm	cm	cm

Experiment 11: Retinal Disparity – Describe your experience

Experiment 12: Visual Nystagmus – Did you diagnose this on your partner?

Experiment 13: Color blindness – Do you have this?

Experiment 14: Afterimage

What did you observe on the wall when you looking away from the bright light bulb?

Red square: _____   Blue square: _____   Yellow square _____

III. Examining the Auditory System

Experiment 15: Rinne's test (N = normal, D = deafness)

Right ear _____   Left ear _____

Experiment 16: Weber's test (N = normal, D = deafness)

Right ear _____   Left ear _____

What happened when one ear was plugged?

Experiment 17: Sound byte - Describe your experience

IV. Examining the Vestibular System

Experiment 18: Vestibular nystagmus

What did you observe after the subject had stopped spinning?

Spinal Reflexes & Electromyogram (EMG)

Study Questions

1. Name and describe the five parts of the reflex arc.
2. Outline the sequence of events between tapping of the patellar ligament and contraction of the thigh muscles.
3. When testing reflexes, what is the most important thing to observe?
4. If the spinal cord is damaged <u>at the level</u> of a muscle reflex, what happens to the reflex? What happens to the reflexes that originate above the area of the spinal cord damage?
5. List the steps of transmission of the action potential from nerve to muscle.
6. List the steps from generation of an action potential in the muscle to contraction in the muscle cell.
7. What happens inside the skeletal muscle cell that causes the twitch to end?
8. Why does a total lack of ATP result in rigor mortis?
9. What is recruitment?
10. In the body, why are muscle cells recruited in motor units (groups) instead of individually?
11. What is summation?
12. Explain the results from the electromyogram (EMG) exercises.
13. What is the agonistic muscle group? What is the antagonistic muscle group?
14. Why does the EMG amplitude increase when one lifts a heavier load? (Lab experiment)
15. What are the following, and how do they relate to muscle contraction?
    a. Alpha motor nerve.
    b. Actin.
    c. Sarcoplasmic reticulum (SR).
    d. Acetylcholine.
    e. Nicotinic receptor.
    f. Excitation–contraction coupling.
    g. Tropomyosin.
    h. Acetylchoninesterase.
    i. Myosin.
    j. Troponin.
    k. T Tubule.

Applied Questions (Answers in Back)

1. How might an EMG be used to teach a patient to relax tense muscles?
2. Tubocurare blocks the nicotinic receptor on the skeletal smooth muscle. This drug is shot into animals in Africa to paralyze them. Why would this drug paralyze the animals?
    a. In the presence of this drug, would the motor nerve still have an action potential?
    b. Would the skeletal muscle still have an action potential?

3. Myasthenia Gravis is a disease in which the nicotinic receptor on the skeletal muscle is destroyed by the patient's own immune system.
   a. What would be the symptoms?
   b. Would this patient have a muscle action potential for every motor neuron action potential? Why or why not?
   c. The treatment for this disease is to give the patient a drug to inhibit acetylcholinesterase. Why would this treatment be helpful?

Introduction

Reflexes

A **reflex** is an automatic involuntary nervous system response to an internal or external stimulus. Reflexes allow the body to maintain a relatively constant internal environment by coordinating functions of multiple organ systems, and allowing a quick response to potential harm. Reflexes utilize negative feedback systems consisting of the following: 1) **Receptor**: a structure at the end of a sensory neuron that receives information, 2) **Afferent pathway**: a **sensory neuron** that carries the information from the receptor to the central nervous system (CNS), 3) **CNS**: the brain and spinal cord that processes/integrates the information from the afferent pathway and relays information on how the body should respond to the initial stimulus, 4) **Efferent pathway**: a **motor neuron** that carries information from the CNS to the effector organ, 5) **Effector organ**: usually a muscle or gland that appropriately responds to the initial stimulus.

Some reflexes are multisynaptic, involving many synapses; while others are **monosynaptic**, involving only one synapse between an afferent and efferent neuron. Certain reflexes occur entirely on one side of the body and are called **ipsilateral reflexes**; while **contralateral reflexes** involve afferent neurons on one side of the body and efferent neurons on the other side of the body. Some reflexes (Ex. eye blink reflex) directly involve the brain, whereas others (Ex. patellar tendon knee-jerk reflex) involve only the spinal cord. The patellar-jerk reflex is an example of a **stretch reflex**. When the patellar tendon is struck with a mallet, tendon stretch (initial stimulus) activates stretch receptors (intrafusal fibers) within the quadriceps muscles of the front thigh. The intrafusal fiber stretch receptors send sensory information via the afferent pathway to the dorsal horn of the spinal cord. Inside the spinal cord, the afferent neurons directly synapse with the efferent (motor) neurons, which exit the ventral root of the spinal cord, and stimulate the quadriceps muscles to extend the lower leg (kick). Leg extension ultimately decreases stretch on both the muscle and tendon preventing the quadriceps from overstretching and potentially tearing. The patellar-jerk reflex is an example of a monosynaptic ipsilateral reflex.

Physicians often test stretch reflexes to determine whether patients have spinal cord damage and/or to localize an area of spinal damage. Strength of the reflex is less important than <u>evenness</u> of response on both sides of the body. If one side of the body responds more strongly than the other side, then the patient may indeed have spinal cord damage or disease. When the spinal cord is damaged, reflexes **at the site of the damage** will be weak or absent. Reflexes **above** a damaged area of the spine (closer to the brain) will likely not be affected. Interestingly, reflexes **below** the damaged area may actually be exaggerated due to the loss of *inhibitory* signals descending from the brain to the spinal cord.

Neuromuscular Junction

In order for voluntary movement to occur (Ex. walking, lifting weights, etc.) the nervous system must command our voluntary/skeletal muscles to contract (as voluntary muscles are physically connected to our bony skeleton via tendons- shortening of skeletal muscle will cause body movement). Indeed, axons of alpha motor neurons exit the ventral horn of the spinal cord to synapse with and command voluntary/skeletal muscles. The **neuromuscular junction** (NJM) is defined as the junction/synapse between the motor neuron (axon terminus) and the innervated skeletal muscle (motor endplate). Often, a single alpha motor neuron axon will branch a few times before making a synapse with individual muscles cells (within same overall muscle, Ex. bicep) A <u>motor unit</u> is defined as: one motor neuron plus all of the individual muscle cells that it makes a synapse with.

To initiate movement, a number of molecular steps must occur at the NMJ to ultimately signal skeletal muscle contraction. First, an action potential initiates within the alpha motor neuron, and is propagated to the axon terminus. This depolarization opens voltage-gated Ca^{++} channels, and extracellular Ca^{++} enters the presynaptic terminus signaling exocytosis of **acetylcholine** (ACh). ACh diffuses across the NMJ synaptic cleft, and binds to nicotinic ACh receptors on the motor endplate of the muscle cell. Binding of ACh to the nicotinic receptor (ligand gated ion channel) triggers conformational changes in the receptor that lead to the channel opening and ion flow. The predominant ion to flow through the nicotinic receptor is Na^+, and the inward drive of Na^+ causes the motor endplate to depolarize- leading to an excitatory endplate potential (EPP). This EPP then spreads along the muscle cell membrane, and brings nearby voltage-gated Na^+ channels to threshold. This triggers an action potential to occur within the muscle cell, and spread throughout the entire muscle plasma membrane. The action potential spreads along the t-tubules, and triggers Ca^{++} release from the sarcoplasmic reticulum (SR). Ca^{++} influx into the cytoplasm signals protein interactions that ultimately cause muscle cell shortening. ACh is degraded by the enzyme, acetylcholinesterase, which prevents ACh from repeatedly stimulating the muscle, and assuring a single muscle contraction/twitch for each motor neuron action potential. The amount of ACh released from an alpha motor neuron is typically very large- always getting the EPP to threshold (in health). Certain poisons/pharmaceuticals target the NMJ. For example, tubocurarine is a poison derived from the bark of certain South American trees that indigenous tribes used to make poison arrows. Interestingly, curare binds to and blocks the nicotinic receptor. By serving as an antagonist to the nicotinic receptor, curare prevents ACh from binding, and blocks NMJ signaling- leading to paralysis and even death of prey exposed to the poison.

Muscle Contraction

Skeletal muscles are long multinucleate cells that extend from one tendon (insertion to bone) to the other tendon. Inside muscle cells are a series of specialized contractile proteins (actin and myosin) that are arranged in a very orderly and repetitive manner throughout the length of the muscle cell. A single unit of this orderly series of actin/myosins is called a **sarcomere**. As sarcomeres align along the length of the muscle cell in a very repetitive fashion- voluntary muscles look "striated" (striped) under the microscope. In order to shorten then, muscle cells signal for their actin/myosin proteins to interact with each other and increase their degree of overlap. As mentioned in the previous section, events at the NMJ lead to Ca^{++} entry into the muscle cell cytoplasm (from the SR). Elevated Ca^{++} levels lead to molecular events that cause muscle cell shortening; this process is called **excitation/contraction coupling**. At rest, the regulatory protein **tropomyosin** is typically oriented across the thin filament protein **actin** inhibiting the motor protein **myosin** from interacting with actin and "cross bridging." During excitation contraction coupling, elevated cytoplasmic Ca^{++} binds to a second regulatory protein called **troponin**. Ca^{++} binding induces a conformational change in troponin, which in turn leads tropomyosin to also change shape. This conformational change in tropomyosin causes it to rotate around the actin filament- exposing the precise binding sites upon actin that myosin is able to interact with. Thus, myosin (using the power of ATP) will cross-bridge with actin and "powerstroke" neighboring actins towards each other- thereby shortening the sarcomere and muscle. Cross-bridge cycling will continue until enough Ca^{++} is pumped back into the SR to cause troponin to resume its original shape- leading tropomyosin to again cover the actin binding sites. ATP is required in two different aspects of muscle contraction- 1) "recocking" of the myosin head group (in preparation for the next powerstroke), and 2) pumping of Ca^{++} back into the SR by the smooth ER Ca^{++} ATPase (SERCA pump). If ATP is not present (ex. death), the actin and myosin bond cannot be broken and muscles become locked in position; this phenomenon is known as rigor mortis. As all the force used to produce a muscle contraction is generated via actin/myosin crossbridge cycling, the only way to affect a given muscle contractile strength is by affecting the amount and efficiency of this cross-bridging. There are three major methods to do this: 1) position the muscle closer to optimal length, 2) increase the action potential frequency along a single neuron via treppe, summation, and tetany, and 3) recruitment- adding more motor units to the contraction.

Electromyogram

The EMG is a measurement of electrical changes on the skin surface that result from action potentials generated in the muscle cells underlying the skin. Thus, a stronger muscle contraction results in an EMG wave of larger amplitude, and a weaker muscle contraction results in an EMG wave of smaller amplitude (**Figure 5.1**).

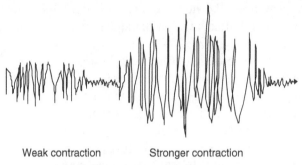

Weak contraction   Stronger contraction

Figure 5.1

Your computer automatically integrates the EMG wave to determine the amplitude. This integration wave gets higher as the EMG amplitude gets larger (**Figure 5.2**).

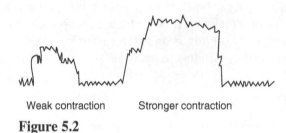

Weak contraction   Stronger contraction

Figure 5.2

Experiment 1: Testing Spinal Reflexes

Procedure

NOTE: It is important that your partner completely relax their muscles during the test.

I. **Patellar-Jerk Reflex** tests for damage to the **femoral nerve**.
1. Have your partner sit on the lab bench with the lower leg hanging over the edge.
2. Strike the patellar ligament just below the knee cap (**Figure 5.3**).
3. Repeat on the other leg.
4. Record your results on the data sheet at the end of this exercise.

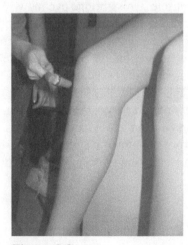

Figure 5.3

II. **Ankle-Jerk Reflex** - tests for damage to the **medial popliteal nerve**.
    1. Have your partner take off his/her shoes and socks then kneel on the lab stool.
    2. Strike the Achilles tendon with the mallet (**Figure 5.4**).
    3. Repeat on the other side.
    4. Record your results on the data sheet at the end of this exercise.

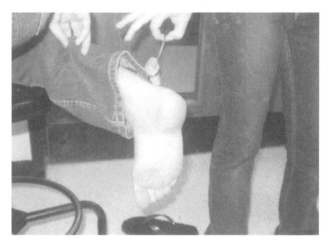

Figure 5.4

III. **Biceps-Jerk Reflex** - tests for damage to the **musculotaneous nerve**.
    1. Rest your partner's elbow in one hand with the thumb of that hand pressing on the biceps tendon.
    2. Using your other hand, strike your thumb with the mallet (**Figure 5.5**). You should observe a slight twitch of the biceps muscle.
    3. Repeat on the other arm.
    4. Record your results on the data sheet at the end of this exercise.

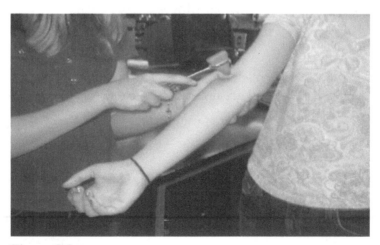

Figure 5.5

IV. **Triceps-Jerk Reflex -** tests for damage to the **radial nerve**.
1. Have your partner face you and place the bend of his/her elbow between your thumb and index finger.
2. With the other hand strike the triceps tendon about 2 inches above the elbow (**Figure 5.6**).
3. Repeat on the other arm.
4. Record your results on the data sheet at the end of this exercise.

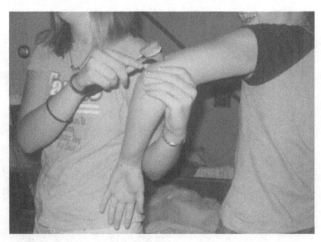

Figure 5.6

Experiment 2: Performing an EMG

Setup

1. One member of your group should perform each of the following functions:
    a. **Director:** Reads the directions.
    b. **Controller:** Runs the computer.
    c. **Subject:** person from whom data are being recorded.
2. Log in to the computer at your station.
3. Turn the MP36 on using the switch at the back of the unit.
4. Double click on the BIOPAC BSL Student Lab 4.1 icon on the desktop.
5. Click on the PRO Lessons tab, select EMG and then click OK.
6. The following three channels should appear:
    *   Channel 1: Force generated on the hand dynamometer.
    *   Channel 40, Integrated EMG: Amplitude of the EMG wave for the digital flexor muscle.
    *   Channel 41, Integrated EMG: Amplitude of the EMG wave for the biceps muscle.

Electrode Connections

1. Using an alcohol swab, scrub the skin surface at each site (**Figures 5.7A and B**) and wipe dry with a paper towel before placing the electrodes.
2. For the digital flexor muscle, measured on channel 2 (Figure 5.7A)
    *   Electrodes:
        *   Place one electrode near the elbow, off center, closer to the ventral (inner) arm.
        *   Place the second on the underside (inside) of wrist about 2 inches up from the hand, near the thumb.
        *   Place the third (ground) on the underside (inside) of the wrist, near the little finger.

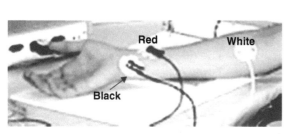

Figure 5.7A

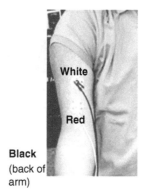

Figure 5.7B

- Leads:
  - Find the set of electrode leads that are connected to channel 2.
  - The pinch connectors work like small clothes pins, but must be placed with the metal clip portion down.
  - Attach the <u>white</u> lead to the electrode closest to the elbow, the <u>red</u> lead to the electrode closest to the thumb and the <u>black</u> lead (ground) to the electrode closest to the little finger
3. For the biceps muscle, measured on channel 3 (Figure 5.7B):
   - Electrodes:
     - Place one electrode over the proximal part of the biceps muscle.
     - Place the second over the distal part of the biceps muscle.
     - Place the third (ground) over the middle of the triceps muscle on the back of the upper arm (not shown).
   - Leads:
     - Find the set of electrode leads that are connected to channel 3.
     - Attach the <u>white</u> lead to the electrode near the top of the biceps muscle (*i.e.*, closest to the shoulder), the <u>red</u> lead to the electrode at the bottom of the biceps muscle and the <u>black</u> lead (ground) to the triceps muscle (not shown).
4. Before recording, note the following:
   - Apply the electrodes at least 3 minutes before recording.
   - Nothing should rub against the electrodes, including the subject's clothing.
   - No pressure should be placed on the electrodes or the baseline will be distorted.

Calibration

1. Click on Start.
   - **Note:** Waves in the three channels may not be apparent until after you autoscale.
2. The subject should relax, holding, but not squeezing, the hand dynamometer in his/her hand for 2 seconds.
3. He/she should then clench the dynamometer as hard as he/she can for 2 seconds and then relax for 2 seconds.
4. Click on Stop.
5. Right click anywhere on the graph window. On the drop down menu that appears, click on Autoscale All Waveforms.
6. Your results should resemble **Figure 5.8**. If they don't, contact your instructor.
7. Save your data (File>Save as) in your section's folder within the Student Data folder on the desktop. Name your file EMG.

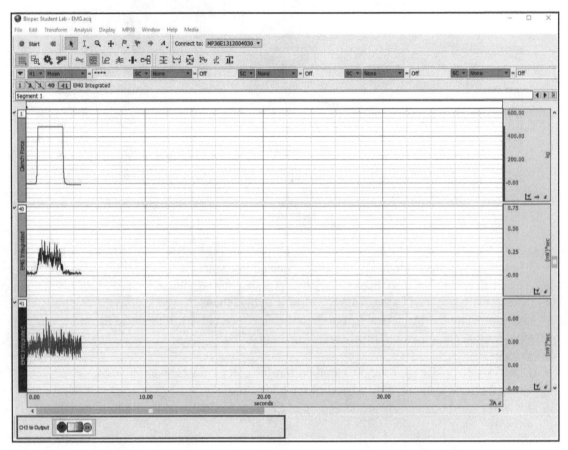

Figure 5.8

I. Strength of Digital Flexor Contraction and EMG Wave Amplitude

Procedure

1. Click on Start. The Controller should run the computer while the director reads the directions to the subject.
2. The subject should squeeze the hand dynamometer weakly (about 20% of his/her original strength) for 3 seconds and then relax for 3 seconds.
3. He/she should then squeeze the hand dynamometer moderately (about 50% of his/her original strength) for 3 seconds and then relax for 3 seconds.
4. Finally, the subject should squeeze the hand dynamometer as strongly as he/she can for 3 seconds and then relax for 3 seconds.
5. The controller should click on Stop.
6. Save your data (File>Save).
7. Print your graph to the printer labelled **TS380 - HP1505N on MAVPRINT**. Make sure that the orientation is Landscape and the Fit Sheet on One page box is selected. Your graph should have three strengths of contraction visible on the screen.

II. Strength of Biceps Contraction (Lifting Weights) and EMG Wave Amplitude

Procedure

1. Turn on the headphones by clicking on the ON button on the Channel 3 to Output in the bar at the bottom of the graph window (**Figure 5.9**).

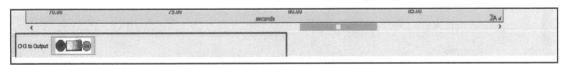

Figure 5.9

2. The subject should place the headphones loosely over his/her ears to hear the EMG activity in the form of noise. The sound may be quite loud. If the sound is uncomfortably loud, contact your instructor.
3. The subject should then stand up, keeping the arm of interest relaxed at his/her side.
4. The controller should then click on Start.
5. The subject should lift his/her hand up slowly, bending the elbow, and the put it down slowly (*i.e.*, perform a bicep curl and then move the arm back to the relaxed position). He/she should then relax the arm for 3 seconds.
6. The director should hand the 2 lb weight to the subject while his/her arm is still in the relaxed state.
7. Once the subject is holding the 2lb weight, he/she should lift the weight up slowly, bending the elbow, and then put it down slowly before returning the weight to the director. He/she should relax the arm for 3 seconds.
8. Repeat steps 6 and 7 with the 5 and 10 lb weights.
9. If the waves are too big and go off the screen, autoscale the data.
10. Save your data (File>Save).

III. Analysis of Data

A. Measuring the Biceps EMG Wave

1. Make channels 1 and 40 invisible (See Exercise 2).
2. Click on the second icon from the left under the Start menu and then on journal in the drop down menu that appears (See Exercise 1, Figure 1.5).
3. Measure the EMG for the empty hand. To do so, highlight the level part of the peak in the Integrated EMG Biceps channel (**Figure 5.10**). Click on Ctrl-M on your keyboard to record your data in the journal.
   NOTE: If your journal has anything other than numbers (*i.e.*, channel labels or units), contact your instructor before going on.
4. Repeat step 3 for the contractions of the biceps using the 2 lb, 5 lb and 10 lb weights. Save your data (File>Save).

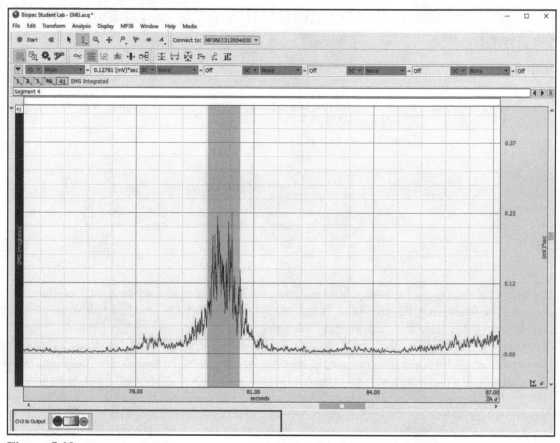

Figure 5.10

B. Graphing the Biceps Data

1. Select all the data in the journal by dragging across the values to highlight them. Click on Ctrl-C on your keyboard to copy the data.
2. Open the Excel Templates folder within the Bio 330 folder on the desktop.
3. Double click on the EMG template to open.
4. In the EMG workbook, select the cell in the B column (B2), directly under the cell labeled EMG (v) (**Figure 5.11**). Click on Ctrl-V on your keyboard to paste the data.
5. Print your graph to the printer labelled TS380 - HP1505N on MAVPRINT. Make sure that the orientation is Landscape and the Fit Sheet on One page box is selected.
6. Save your data (File>Save).
7. Turn in the printout of your graph (be sure to put all group members' names on the graph) along with the data sheet at the end of this exercise to your instructor at the end of your lab session.
8. Close the MP36 graph window and turn the Biopac off via the switch in back of the MP36 unit.

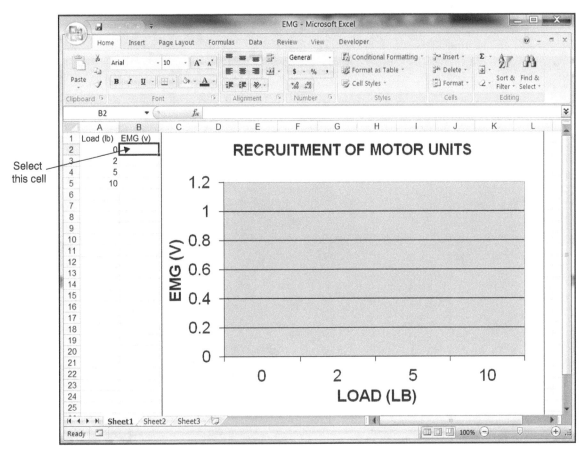

Select
this cell

Figure 5.11

Spinal Reflexes & Electromyogram Data Sheet

Group Number: _____   Name: _____

Experiment 1: Spinal Reflexes

Reflexes (results – check ($\sqrt{}$) the correct box)

Reflex	Same on both sides	Stronger on one side
Patellar-jerk reflex		
Ankle-jerk reflex		
Biceps-jerk reflex		
Triceps-jerk reflex		

Experiment 2:

Muscle contractions with hand dynamometer

What is being measured with an EMG? Explain what is happening in the muscle to produce the EMG.

Using your graph, describe and explain the relationship between the dynamometer wave on channel 1 and the EMG wave on the second channel (Channel 40). Specifically, why do they increase and decrease together?

Muscle output and increasing resistance

Describe the relationship between the weight you lifted and the EMG wave. What is occurring physiologically to produce this relationship?

Based on your results, <u>predict</u> what you would expect the EMG to look like if you picked up your cell phone. What if you picked up your backpack containing your physiology textbook?

///////// EXERCISE
6

Skeletal Muscle

Study Questions

1. Explain why a larger stimulus voltage results in a stronger muscle contraction. What is the name for this property?
2. In your body, how are additional muscle cells stimulated to contract?
3. Why does a greater than maximal (supramaximal) stimulus voltage to a skeletal muscle <u>not</u> result in a stronger contraction than the maximal stimulus voltage?
4. What is <u>treppe</u>? What is believed to be the mechanism for this stronger contraction?
5. Why can a second contraction occur in a skeletal muscle before the first contraction has ended? Why doesn't the refractory period prevent this?
6. What is <u>summation</u>? How did we produce this in our frog muscle?
   a. How are the series elastic elements involved in the stronger contractions produced during summation?
   b. How is Ca^{++} involved in the stronger contractions produced during summation?
7. What is tetany? How did we produce this in our frog muscle?
8. Using actin and myosin overlap in the sarcomere, describe the effect of stretch on the strength of a muscle contraction, and explain the reasons for this effect.
   a. How did we demonstrate this property in lab?
9. Which of the three muscle properties (recruitment, summation/tetany, and length–tension) were demonstrated in each of the following exercises?
   a. Increasing stimulus voltage.
   b. Stretching the muscle.
   c. Increasing frequency of stimulation.

Introduction

For appropriate coordinated movement to occur, the body needs to be able to adjust contractile strength of a given set of muscles (think picking up a feather versus a bowling ball). There are several mechanisms that the body uses to affect the strength of any given muscle contraction. All of these mechanisms affect either the amount or efficiency of actin/myosin cross-bridging (more cross-bridging = more strength produced).

Recruitment is the adding of more motor units to make a stronger muscle contraction. Remember, a motor unit is a single motor neuron plus the individual muscle cells that it innervates. Since each large muscle (Ex. Bicep) is composed of many different parallel muscle fibers, the body can affect the strength of contraction of a given overall muscle (Ex. Bicep) by affecting the number of motor units/muscle fibers involved in that particular contraction. Once an individual muscle cell is brought to threshold, that cell completes an "all-or-none" twitch. Therefore, a whole muscle contraction that uses only a few of the available cells will be weaker than a contraction that uses most or all of the individual cells in the muscle. In this way, the body can <u>recruit</u> additional muscle cells when a stronger contraction is needed- by adding in more motor units to the contraction. As additional motor neurons are stimulated, more motor units/cells will be added to the contraction - and more cross-bridging/strength will be produced. A skeletal muscle's contractile elements must shorten for a period of time to stretch the **series elastic elements**, which are composed of elastic-like connective tissue and are located on the ends of the sarcomeres. Once these elements have developed adequate tension, the whole muscle shortens- and movement is produced.

With any <u>single</u> action potential, a skeletal muscle cell releases the same amount of Ca^{++}, which uncovers the same number of cross-bridge sites on the actin molecule. Thus, any single, well-spaced muscle cell stimulation produces a twitch of the same strength.

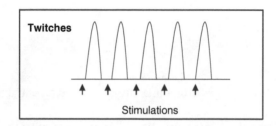

However, a skeletal muscle that is stimulated with a greater **<u>frequency</u>** (stimulations are closer together) will have stronger contractions due to **Treppe**, **Summation**, and **Tetany**. **Treppe** occurs when stimulations are close enough together to produce a stronger contraction, but far enough apart to allow the muscle tension to return completely to baseline.

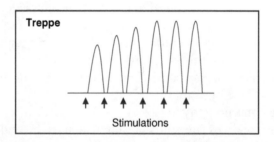

Potential explanations for the "Treppe effect" include not enough time has elapsed between twitches to pump all the Ca^{++} back into the SR, therefore causing a buildup of Ca^{++}, and more cross-bridges to be uncovered. An increase in muscle temperature due to increased activity may also improve speed of cross-bridge cycling during Treppe.

A skeletal muscle cell that is stimulated with even higher frequency of action potentials- when the subsequent action potentials stimulate contractions even *before* the muscle has completely relaxed from the first contraction, will have a second <u>stronger</u> twitch, because the second contraction is *stacked on* the first in a process called **<u>Summation</u>**. Please note that the refractory period of the muscle action potential **does not** interfere with stimulation of a subsequent action potential because the action potential is quite short in duration relative to the actual contraction events (Ca^{++} entry, cross-bridge cycling, etc. . .).

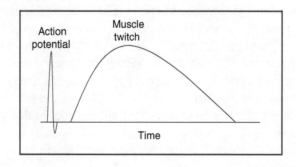

Contractions during summation increase in strength because the second contraction begins from a partially contracted state. Thus, the second contraction occurs <u>before</u> the series elastic elements have completely relaxed, and less time is spent stretching the elements (more time is spent shortening) so the muscle can shorten further - developing even greater tension. As in treppe, elevated Ca^{++} levels are also likely involved ensuring maximum troponin bound, and therefore maximum cross-bridge cycling produced.

When the muscle is stimulated with a series of stimulations so close together that <u>no relaxation</u> occurs between contractions, the muscle is in **Tetany**, and has achieved maximum contractile force.

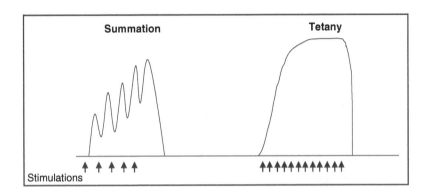

One final way to affect the strength produced from a given skeletal muscle is the **length tension relationship**. Indeed, the resting length of a muscle (degree of stretch) has a marked effect on its contraction strength. A muscle that is <u>too short</u> at the time of contraction has a weak contraction, because the degree of overlap of the actin and myosin filaments is too great, and therefore the opposing actins interfere with each other as they slide towards each other during the contraction. Conversely, a muscle that is stretched <u>too long</u> has a weaker contraction because there is too little overlap of the actin and myosin filaments. Therefore, a muscle at the optimal length contracts the strongest because the actin and myosin overlap is ideal for optimal for cross-bridge interaction.

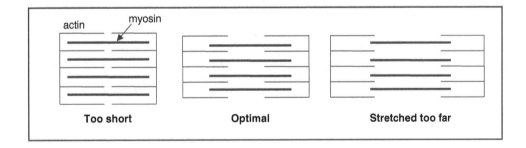

Objectives of Experiments 1-3

1. To observe and record recruitment in a frog skeletal muscle by increasing the stimulus strength (voltage).
2. To observe and record summation and tetany by increasing the frequency of stimulation.
3. To observe and record the length–tension relationship of skeletal muscle by recording the contraction strength during changes in resting (starting) muscle length.

Setup

1. Log in to the computer at your station.
2. Turn the MP36 on using the switch at the back of the unit.
3. Double click on the BIOPAC BSL Student Lab 4.1 icon on the desktop.
4. Click on the PRO Lessons tab, select Frog Muscle and then click OK.

5. Two channels, channel 1, which records the voltage of the stimulus applied to the muscle, and channel 2 that records the force of contraction, should be apparent.

Calibration

1. Be sure the S-hook is on the 1000-gram weight eye (farthest from the end) of the muscle transducer (**Figure 6.1**).

Figure 6.1

2. Hang the two weights on the S-hook as shown in **Figure 6.2**.

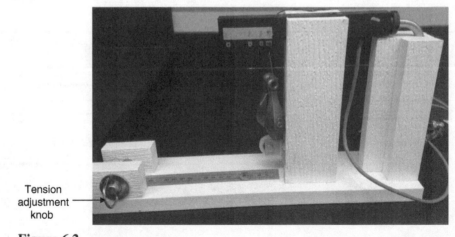

Tension
adjustment
knob

Figure 6.2

3. Select the small wrench icon to the right side of channel 2 in the grams area (see circle, **Figure 6.3**).

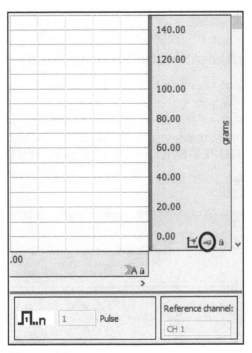

Figure 6.3

4. The following scaling analog channel dialog box should appear (**Figure 6.4**).

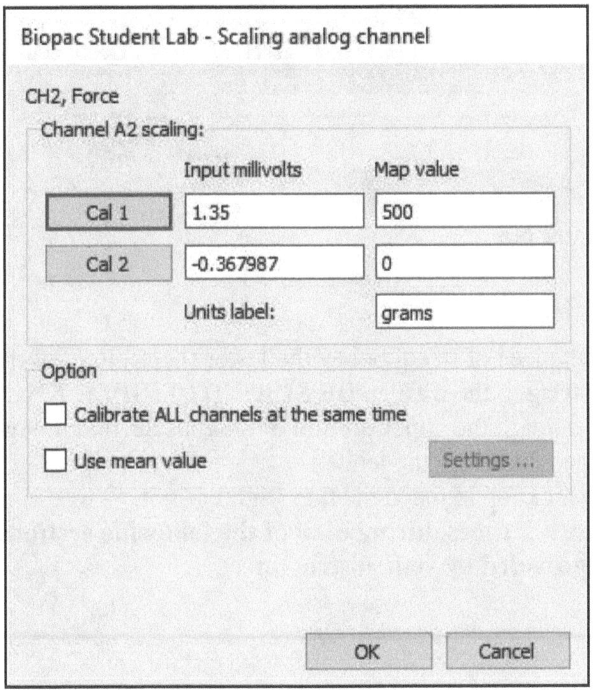

Figure 6.4

5. Click on Cal 1. This calibrates the known weight to 500 grams.
6. Remove the weights, leaving only the S hook hanging from the force transducer.
7. Click on Cal 2. This calibrates the zero force.
8. Click on OK.

Muscle set up (done only after calibration is completed)

1. Turn the tension adjustment knob (arrow in Figure 6.2) to move the washer to 50 mm on the ruler attached to the muscle apparatus at your station.
2. Tie one end of a 6-inch string around the the muscle, above the knee joint. Using the same end of the string, tie it around the S-hook, making sure that the S-hook is as close as possible to the muscle (**Figure 6.5**). **BE SURE TO TRIPLE KNOT AT EACH LOCATION!!**

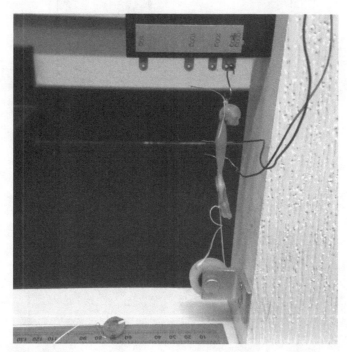

Figure 6.5

3. Tie one end of the 12-inch piece of string below the lower (hock) joint of the muscle.
4. Tie the other end of the string to the washer. **BE SURE TO TRIPLE KNOT AT EACH LOCATION!!**
5. Push one electrode wire through the upper portion of the muscle, and the second through the lower end of the muscle. Do not place them into the tendon.
6. Turn the tension adjustment knob to make the line slightly tight.
7. **Wet your muscle at least 1–2 times during each of the following sections with the pipette filled with Ringer's solution provided by your instructor.**

IMPORTANT NOTES

- **Before you begin the experiments, read through the entire procedure.**
- **The computer will continue to record unless you press Stop. Do not stop unless you need to.**
- **Have one person watch the muscle to see a twitch, and another watch the screen for a recorded response.**

Experiment 1: Recruitment

In this section of the lab, you will bring muscle cells to threshold by electrical stimulation. A small stimulus voltage will bring only a few muscle cells to threshold, and will create a weak muscle twitch. A larger stimulus voltage will stimulate more muscle cells and will create a stronger muscle twitch. As voltage is increased, individual cells are added to the contraction, a process called **recruitment**. Note, a stronger twitch creates a taller contraction wave on the computer.

Procedure

1. Click on Start.
2. View the line tension on channel 2. If need be, adjust the tension to approximately 30 grams (**Figure 6.6**). Click on Stop when you have finished.

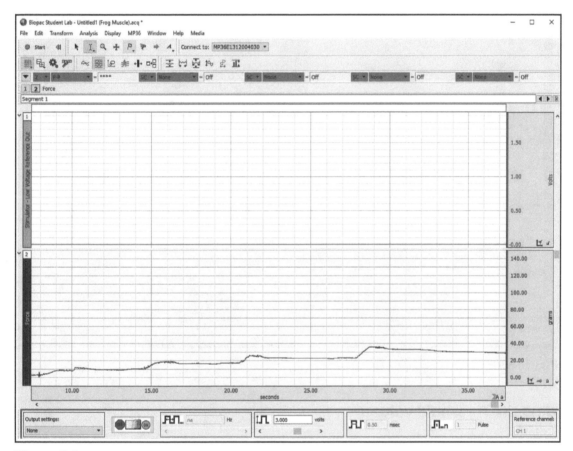

Figure 6.6

3. Let the setup sit for a minute.
4. Click on Start, and if needed, adjust the tension again until the force on channel 2 is 30 grams. Click on Stop.
5. Save your data (File>Save as) in your section's folder within the Student Data folder on the desktop. Name your file Frog Muscle.

I. Testing the Muscle

1. Set the stimulator at the bottom of your graph window (See Exercise 3, Figure 3.1) to 3 volts.
2. Click on Start.
3. Stimulate the muscle by clicking the ON button in the stimulator bar at the bottom of the window (circle, **Figure 6.7**).

4. You should see stimulus on channel 1, and a muscle twitch recording on channel 2 (Figure 6.7).
5. Click on Stop
   Note:
   • If your twitch on channel 2 is bigger than the graph (not all the twitch is showing), autoscale your data.
   • If your data does not look like Figure 6.7, contact your instructor.
6. Save your data (File>Save).

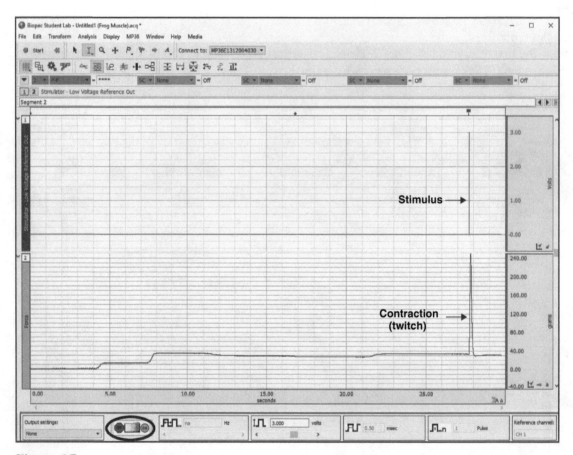

Figure 6.7

II. Finding the Threshold Voltage

1. Decrease the stimulator voltage to 0.4 volts. Press Enter.
2. Click on Start.
3. Click the ON button in the stimulator bar. A stimulus should be visible on channel 1 but no response (flat line) should be apparent on channel 2. If you do see a response on channel 2, reduce your stimulus voltage to 0.1 volts and then click ON again. You should have no response on channel 2.
4. Increase the stimulus strength by 0.05 volts (to 0.45 or to 0.15) by clicking once on the right arrow under the voltage setting in the stimulator window. Click on the ON button again.
5. Continue to increase voltage by 0.05 volts and stimulate once after each increase, until a small twitch is visible on channel 2 (**Figure 6.8**).
6. Press Stop.
7. Insert a marker event above the stimulus by right clicking in the bar below the event bar. In the drop-down menu that appears, click on Insert New Event.
8. Right click on the marker triangle and click Edit event. Highlight and delete any writing in the marker text box, and type: Threshold (See Exercise 3, Figure 3.2).

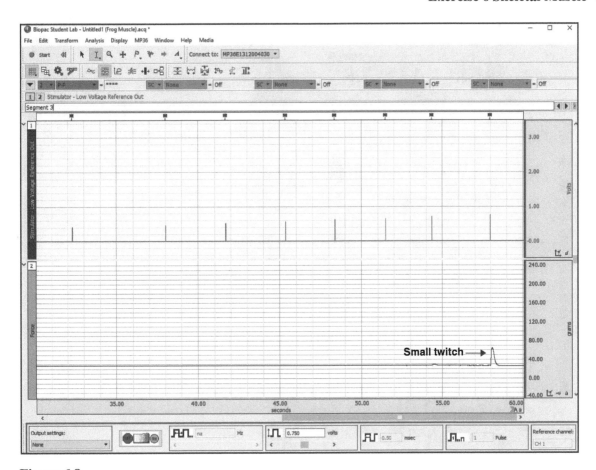

Figure 6.8

III. Finding the Maximal Voltage

1. Increase your voltage by 0.05.
2. Click on Start and then click on ON in the stimulator bar. A taller twitch should appear on channel 2.
3. Continue increasing by 0.05 volts and stimulating via the ON button, until your twitch amplitude does not increase over the previous twitch.
4. To be certain that you have stimulated all the muscle cells, increase your voltage 0.20 volts and then stimulate.
5. Repeat the 0.20 volt increases (and stimulate each time) until your twitch amplitude does not increase over the previous twitch. You should have two twitches of the same size showing on Channel 2.
6. Do one more increase of 0.20 volts and stimulate before clicking on Stop. Your last three twitches should be equal in amplitude, or may decrease slightly due to fatigue.
7. Scroll back using the scroll bar at the bottom of the graph window. Find the last twitch that increased in amplitude when compared to the previous twitch (**Figure 6.9**). This is the Maximal Response.
8. Insert a marker event above the stimulus by right clicking in the bar below the event bar. In the drop-down menu that appears, click on Insert New Event.
9. Right click on the marker triangle and click Edit event. Highlight and delete any writing in the marker text box, and type: Maximal (See Exercise 3, Figure 3.2).
10. Save your data (File>Save).

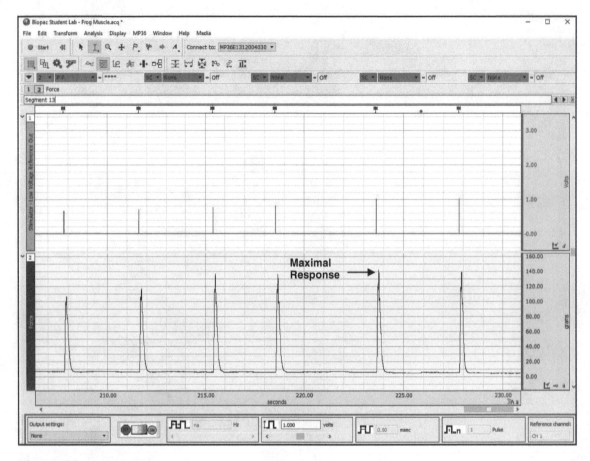

Figure 6.9

Experiment 2: Treppe, Summation and Tetany

The following exercise will examine the strength of a muscle contraction when the muscle when the muscle is repetitively stimulated with multiple action potentials.

Procedure

1. Insert a marker event above the stimulus by right clicking in the bar below the event bar. In the drop-down menu that appears, click on Insert New Event.
2. Right click on the marker triangle and click Edit event. Highlight and delete any writing in the marker text box, and type: Summation/tetany (See Exercise 3, Figure 3.2).
3. Adjust your time scale to span 60 milliseconds (See Exercise 3, Experiment 2, steps 4a - c).
4. Resting tension should be adjusted to approximately 40 grams by turning the tension adjustment knob (arrow in Figure 6.2).
5. In the Output settings box on the lower left part of the graph window (in the stimulator control bar), click on the None and select Sum in the drop down menu (circle, **Figure 6.10**).
6. The pulse rate should change to 1.00 Hz (arrow, Figure 6.10).and the voltage to 3.00 volts

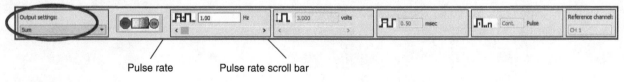

Pulse rate   Pulse rate scroll bar

Figure 6.10

7. Click on Start. **LEAVE THE COMPUTER IN THE RECORDING MODE UNTIL YOU FINISH THIS SECTION!!**
8. Click ON in the stimulator bar. Wait 3 for seconds and then click OFF.
9. Let your muscle completely relax for 10 seconds to prevent muscle fatigue. Use the time scale on the X-axis (along the bottom of the graph) to estimate when 10 seconds have past.
10. Click once on the right arrow of the pulse rate scroll bar on the stimulator bar at the bottom of the graph window (Figure 6.10). This should increase the pulse rate to 2 Hz. Don't worry if your Output setting goes back to None.
11. Repeat steps 8-10 until you think you are close to complete tetany.
12. When you are close to tetany, increase your pulse rate by 5 Hz (5 clicks on the arrow) and stimulate for 3 seconds. If complete tetany has not been reached, repeat this step. Your data should resemble **Figure 6.11**.

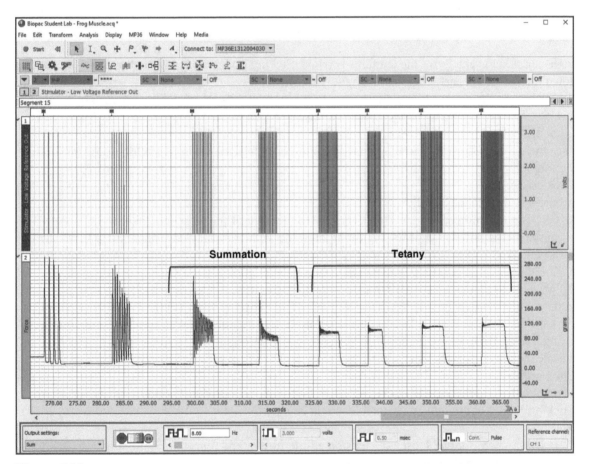

Figure 6.11

13. Press Off in the stimulator window, then press Stop in the graph window. **BE SURE TO PRESS OFF IN THE STIMULATOR WINDOW OR THE MUSCLE WILL CONTINUE TO BE STIMU-LATED AND WILL FATIGUE!!**
14. Save your data (File>Save).
15. If your entire summation/tetany section is visible on your computer screen, print your data to the printer labelled **TS380 - HP1505N on MAVPRINT.** Make sure that the orientation is Landscape and the Fit Sheet on One page box is selected.
16. If not, change your time scale before printing. To do so,
    a. Note the time in seconds, at the very end of your data.
    b. Using the scroll bar at the bottom of the graph window, scroll back to the beginning of your summa-tion/tetany section.

    c. Click on the seconds scale at the bottom of the graph window

    d. A horizontal axis scale window should appear. Highlight the number in the box marked end and type in the seconds value you just noted. This extends your time scale from the beginning of the summation/tetany section to the end of your data.

    e. Click OK.

17. Turn in your printout (See Figure 6.11 - be sure to put all group members' names on the graph) along with the data sheet at the end of this exercise to your instructor at the end of your lab session.

Experiment 3: Length Tension Relationship

The following exercise will examine the strength of a single muscle twitch when the muscle is stretched to different starting lengths.

Procedure

1. Insert a marker event above the stimulus by right clicking in the bar below the event bar. In the drop-down menu that appears, click on Insert New Event.

2. Right click on the marker triangle and click Edit event. Highlight and delete any writing in the marker text box, and type: Length-Tension (See Exercise 3, Figure 3.2).

3. Click on Start.

4. Turn the tension adjustment knob (arrow in Figure 6.2) until the string is slightly loose (at or near 0 grams).

5. Click on Stop.

6. Adjust the force range as follows:

    a. Click on the numbers on the *Y*-axis (grams) to the right of the graph on channel 2.

    b. In the dialog box that appears, highlight the value in the box labeled Upper and type 400 and in the box marked Lower, - 10 grams (minus 10 grams).

    c. Click OK.

7. Adjust your time scale to span 30 milliseconds (See Exercise 3, Experiment 2, steps 4a - c).

8. In the Output settings box on the lower left part of the graph window (in the stimulator control bar), click on the None and select L-T (length-tension) in the drop down menu (circle, Figure 6.10).

9. Using a ruler, measure the length of the muscle in millimeters (mm) from the top to the bottom of the muscle. **Starting muscle length** = _____ 52 _____ **mm**

10. Click on Start.

11. Stimulate the muscle once by clicking the ON button in the stimulator window.

12. Click on Stop. Your twitch on channel 2 should be small.

    Note:

    • If your twitch is large, loosen the tension a little further and repeat your stimulus.

    • If no twitch is visible, slightly <u>tighten</u> your muscle tension and repeat steps 10-12.

13. Click on Start. **LEAVE THE COMPUTER RUNNING THROUGHOUT THIS SECTION!!**

14. Stretch the muscle slightly by turning the tension knob so the marker on the ruler moves two lines (2 mm) farther from the muscle. The muscle should stretch 2 mm, making its current length 2 mm greater than the original length (original length plus 2 mm).

15. Wait one second for a baseline (a horizontal line) to show on the computer at the new level of tension (**Figure 6.12**).

16. Repeat steps 14 and 15 until the twitch becomes very small or disappears (**Figure 6.13**).

    Note: As stretch progresses, you may have to Autoscale to bring your contraction line into view.

17. Click on Stop.

18. Save your data (File>Save).

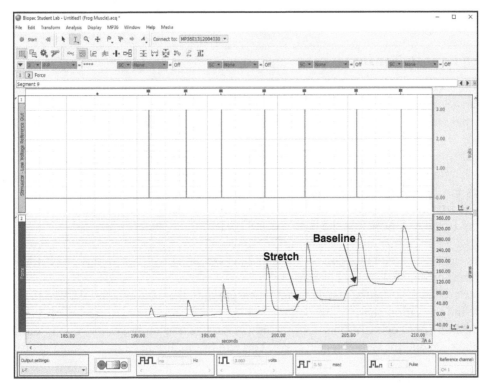

Figure 6.12

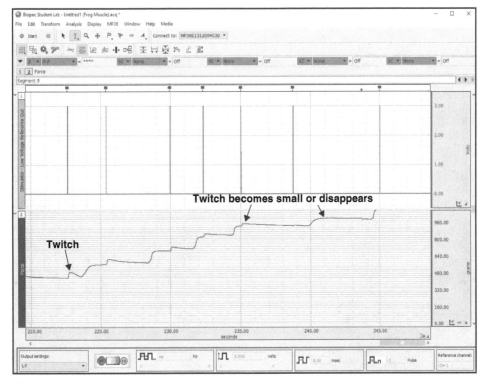

Figure 6.13

Data Analysis

I. Measuring Length-Tension

1. Click on the second icon from the left under the Start menu and then on journal in the drop down menu that appears (See Exercise 1, Figure 1.5).
2. In the journal window that appears at the bottom of the graph window, type Length-Tension and hit Enter.
3. Adjust your time scale to span 10 milliseconds (See Exercise 3, Experiment 2, steps 4a - c)
4. Find your length-tension marker using the left and right arrows (◄►) on the right hand of the screen.
5. Right click in Channel 2 and select Autoscale waveforms.
6. Measure the first muscle twitch of your length-tension section. To do so, scroll to the right and then click on the I-icon to drag across the baseline and peak of the twitch (**Figure 6.14**).
   Note:
   - Be sure each twitch you measure on channel 2 is associated with a stimulus on channel 1 as some deflections on channel 2 may be "noise."
   - If your baseline drops at the end of twitch, measure only the upsweep and peak of the twitch. Do not include the relaxation phase of the contraction.
7. Press Ctrl-M on your keyboard. This will record the muscle tension in grams in your journal.
8. Scroll to the right and measure the next wave. You will eventually need to autoscale to bring the graph into view.
9. Repeat for all the waves in the Length–Tension section.
10. Save your data (File>Save).

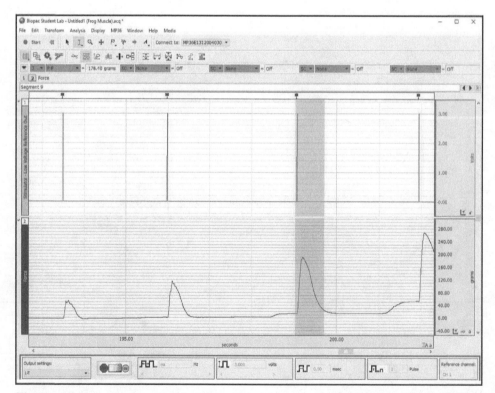

Figure 6.14

II. Graphing Length–Tension

1. Select all the length-tension data in the journal by dragging across the values to highlight them. Click on Ctrl-C on your keyboard to copy the data.
2. Open the Excel Templates folder within the Bio 330 folder on the desktop.
3. Double click on the Length-Tension template to open.
4. In the Length-Tension workbook, select the blank cell in column A (cell A2), just beneath the Length label, but above number 2. Type in your starting muscle length (in mm) you recorded in step 9 of Experiment 3. The computer will automatically enter the remaining lengths.
5. Select the cell in the B column (B2), directly under the cell labeled Tension (g). Click on Ctrl-V on your keyboard to paste the data.
   Note:
   • If your Tension column is longer than your Length column, select the last number in the length column and drag directly down to highlight the remaining cells until you have reached the bottom of the Tension data. Click Cntrl-D to fill down.
   • If your Length column is longer than your Tension column, select and delete the extra numbers.
6. Print your graph to the printer labelled TS380 - HP1505N on MAVPRINT. Make sure that the orientation is Landscape and the Fit Sheet on One page box is selected.
7. Turn in the printout of your graph (be sure to put all group members' names on the graph) to your instructor at the end of your lab session.

III. Measuring Recruitment

1. Click in the journal under the Length–Tension data and type Recruitment. Hit Enter.
2. Adjust the measuring settings at the top of the graph window. This will determine what the computer measures as you highlight your stimulus wave or graph twitches. To do so:
   a. In the first set of measurement settings (**Figure 6.15**), click on the 2 and select CH1, Stimulator- Low Voltage Reference Out from the drop down list. This tells the computer to measure the maximum point on the stimulus wave on channel 1, in the area you will highlight.
   b. In the menu next to the one above, click on the p-p and select max from the drop down list. This tells the computer to measure the maximum point on the stimulus wave on channel 1, in the area you will highlight.
   c. In the next set of two measuring settings (to the right of the first set), click on the SC and change it to CH2, and then on Force and change the none to p-p. The p-p setting tells the computer to subtract the smallest value (baseline) from the highest value (peak).

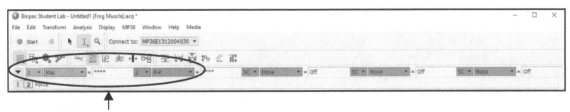

1ˢᵗ and 2ⁿᵈ set of measurement settings

Figure 6.15

3. Find your threshold marker using the left and right arrows (◄►) on the right hand of the screen.
4. Adjust your time scale to span 30 milliseconds (See Exercise 3, Experiment 2, steps 4a - c)
5. Click on the I-icon to drag across your muscle twitch so that your highlighted area includes the stimulus signal in Channel 1 and the entire twitch in Channel 2 (**Figure 6.16**).
6. Press Ctrl-M on your keyboard. Your data should appear in two columns: the first column is stimulus strength in volts and the second is contraction strength in grams.
7. Scroll to the right and find the next stimulus and twitch and repeat steps 5 and 6.

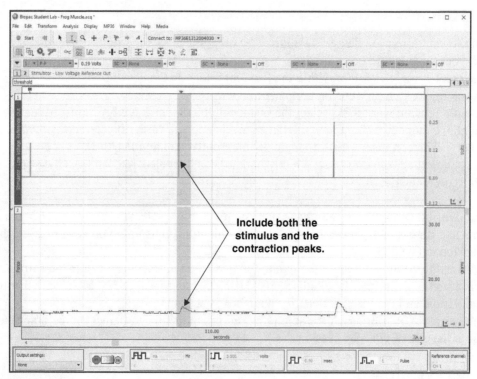

Include both the stimulus and the contraction peaks.

Figure 6.16

8. Measure all stimuli/twitches including two stimulations/twitches past your maximal marker.
9. Save your data (File>Save).

IV. Graphing Recruitment

1. Highlight only the data in the Recruitment section of the journal (do not include the title) and hit Ctrl-C to copy.
2. Open the Excel Templates folder within the Bio 330 folder on the desktop.
3. Double click on the Muscle Recruitment template to open.
4. In the Muscle Recruitment workbook, click on the cell just under the cell labeled Voltage in column A.
5. Click on the arrow under Paste, select Paste Special
6. In the Paste Special Window, choose Unicode Text and click OK.
7. Two columns of data should appear on the sheet, and a bar graph should appear in the graph window. If your data extends beyond Cell 25 (the boxed area), click on the graph and drag the edge of the highlighted boxes down to where your data ends. The graph should adjust automatically.
8. Print your graph to the printer labelled TS380 - HP1505N on MAVPRINT. Make sure that the orientation is Landscape and the Fit Sheet on One page box is selected.
9. Turn in the printout of your graph (be sure to put all group members' names on the graph) along with the data sheet at the end of this exercise to your instructor at the end of your lab session
10. Close the MP36 graph window and turn the Biopac off via the switch in back of the MP36 unit.
11. Discard your muscle, string, gloves and paper towels in the garbage. **DO NOT THROW AWAY THE PIPETTES OR S HOOKS!!**

Skeletal Muscle Data Sheet

Group Number: _____   Name: _____

Observation: What happened to the muscle when you electrically stimulated it? Why?

Experiment 1: Recruitment

How were more motor units recruited during the experiment?

Explain the values on the *x*-axis and the y-axis. (What measurement is graphed in each?)

On your graph, mark the maximal response.

Explain why the bars on the graph level off as you increased the voltage. What occurred in the muscle to produce this response?

Experiment 2: Treppe, Summation and Tetany

How was summation achieved during this experiment?

On your graph, mark the areas in which summation and tetany occurred.

Is any recruitment occurring during this exercise? Why or why not?

Experiment 3: Length-Tension Relationship

How did you determine the optimal length of the muscle?

On your graph, mark the optimal length of your muscle, mark where the muscle is too short and mark where the muscle is stretched too far (too long).

Explain why the tension increases, and then decreases with stretch. What happens within the sarcomere (actin and myosin) that causes this relationship?

Critical thinking question

Explain how you could use all three of these skeletal muscle properties at the same time to allow you to pick up a heavy object. What would be occurring within your body to produce each property?

Electrocardiogram (ECG)

Study Questions

1. Why is it beneficial for the atrioventricular (AV) node to slow conduction between the atria and ventricle?
2. What is the intrinsic depolarization rate of the sinoatrial (SA) node? The AV node? The Purkinje system?
3. Why is the SA node the pacemaker of the heart, even though other areas are capable of automaticity?
4. What happens if the SA node fails? What area usually takes over as pacemaker?
5. Identify the various waves of the electrocardiogram (ECG) tracing and explain what each wave is associated with.
6. During the PR interval (beginning of P wave to the beginning of the QRS complex), where is the depolarization wave in the heart?
7. What is the difference between atrial flutter and atrial fibrillation?
8. Why is the ventricular contraction rate slower than the atrial rate during flutter?
9. What is an AV block? Compare first-, second-, and third-degree AV node block.
   a. Which has the same contraction rate in the ventricles as in the atria?
   b. Which has (have) a slower ventricular rate than atrial rate?
   c. Which has a very slow, but regular ventricular beat?
   d. In a grade III AV block, where is the atrial rate set? Where is the ventricular rate set?
10. What are premature ventricular contractions (PVCs)? Why is there no P-wave preceding this beat?
    a. What is different about the QRS complex?
    b. Is the T-wave normal or abnormal?
    c. Why is this beat followed by a <u>compensatory pause</u>?
    d. Why might this contraction be weak?
    e. Why would the contraction that follows this beat be stronger than usual?
11. What is ventricular tachycardia?
12. Why is ventricular fibrillation such a serious condition?
13. What disorder might result in a prolonged PR interval?
14. What might cause a prolonged QRS complex?
15. What problem can shift the ST segment off the baseline?
16. Know the cause and appearance of the following arrhythmias.
    a. Atrial tachycardia
    b. Atrial fibrillation
    c. Atrial flutter
    d. Second-degree AV node block
    e. First-degree AV node block
    f. Third-degree AV node block
    g. PVC
    h. Ventricular tachycardia
    i. Ventricular fibrillation.

Applied Questions (Answers in Back)

1. Angela Angina had a heart attack and died almost instantly. Attempts by paramedics to revive her were unsuccessful. What arrhythmia did she most likely have?
2. Rodney Rapid has been diagnosed with ventricular tachycardia.
   a. What is setting his rapid ventricular rate? What would his rate of atrial contractions be?
   b. What would you predict about his end diastolic volume (EDV) or stroke volume (SV)? Explain.
3. During atrial flutter, the ventricle does not beat as rapidly as the atria?
   a. Why is this so?
   b. Is the AV node diseased in these patients?

Introduction

An **electrocardiogram (ECG)** is the measurement of electrical activity in the heart that underlies heart muscle contraction. Because the heart undergoes significant and coordinated electrical activity during every contraction, ECG measurements can detect this electricity by placing electrodes onto the skin at disparate locations of the body surrounding the heart. At least three electrodes (or leads) are typically used to perform an ECG measurement- one on each arm and a third on the left leg- to form an approximate triangle around the heart (denominated "Einthoven's triangle"). Each of these electrode sets measures the electrical difference that exists between two points on the body, and this electrical difference changes in a very consistent way as the heart undergoes its routine electrical depolarizations/repolarizations. Therefore, ECG recordings result in a very consistent pattern of waves (please see normal ECG racing below). Because the measuring points differ slightly from each other, the shape and size of the ECG waves can be slightly different between each set of leads. Furthermore, ECG tracings are a powerful investigative tool that are routinely used in the medical clinic to help diagnose heart abnormalities in patients.

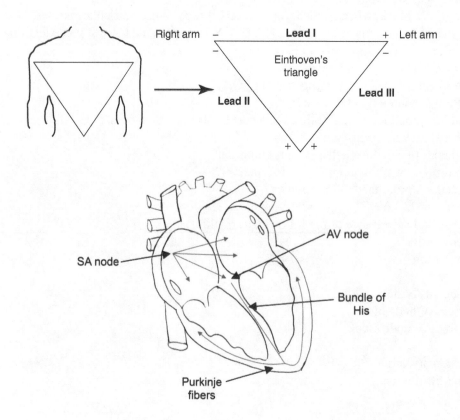

Depolarization of the heart normally begins at the **SA node** (a small population of cells in the upper right atrium that express special "funny" Na$^+$ channels), and spreads throughout both atria via gap channels, and then through the **AV node** and finally onto the ventricles. The AV node's primary job is to delay the electrical conduction somewhat- allowing the atria to depolarize and contract before the ventricles- ensuring blood is "pumped" in the proper direction. Once in the ventricles, the electrical conduction travels through special conduction cells as follows: **Bundle of His** → **bundle branches** → **Purkinje fibers** → contracting cells. The SA node is normally the pacemaker of the heart because its natural depolarization rate is the fastest within the heart (~100 beats per minute). It should be noted that other areas of the heart can also "spontaneously" depolarize (bring themselves to threshold), although their rates are typically slower than the SA node (AV node rate: 40–60 beats per minute, Purkinje fiber rate: 20–40 beats per minute), and therefore they are seldom used as the "pacemaker" in a normal heart. Indeed, any area of the heart that gets to threshold first (before any other area) will bring all heart cells to threshold- since heart muscle cells are interconnected via gap junctions. If an area outside the SA node reaches threshold first, an **arrhythmia**, or abnormal heart rhythm will result. If a single incident occurs, the contraction (beat) is known as a **premature contraction**. If the depolarizations are repeated, a **tachycardia** (fast heart rate) occurs. The "atypical cell" source of the depolarizations is then known as an *ectopic* pacemaker. Thus, the area with the *fastest* depolarization rate always sets the heart rate. If the SA node fails (no longer gets to threshold), the next fastest area of the heart takes over as pacemaker- usually the AV node.

Normal ECG Tracing

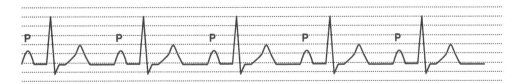

The first wave of an ECG tracing, the **P-wave**, is the result of the depolarization spreading through both atria. The wave is normally positive (upward) on lead II. If an **ectopic** source (a place *other than* the SA node) gets to threshold before the SA node, the depolarization will spread from cell to cell through the entire atria (using gap junctions). When the depolarization of the atria occurs in an abnormal direction, the **P-wave** appears altered, or inverted.

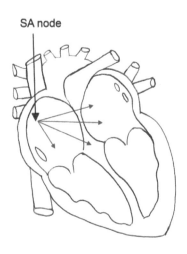

SA node

Normal depolarization of the atria produces a normal P-wave

Ectopic source

If the depolarization is in the reverse direction, the P-wave is inverted

The **QRS complex** results from the depolarization spreading through the ventricles. The R-wave is positive on lead II if the depolarization of the ventricles is normal. The **T-wave** that follows the QRS is the result of repolarization of the ventricles.

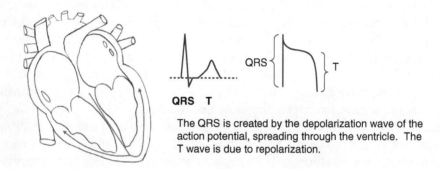

The QRS is created by the depolarization wave of the action potential, spreading through the ventricle. The T wave is due to repolarization.

If an **ectopic source** in the ventricle reaches threshold <u>before</u> the SA node depolarization reaches the ventricle, the **ectopic** depolarization will spread through the ventricle (via gap junctions) in an <u>abnormal</u> direction, and therefore the <u>QRS complex</u> will also be abnormal. If *depolarization* of the ventricle is abnormal, then *repolarization* will also be abnormal, and thus the <u>T-wave</u> will be altered as well. **No P-wave** is visible *before* the abnormal QRS because the ventricular depolarization did not originate in the SA node, and therefore was not preceded by a P-wave.

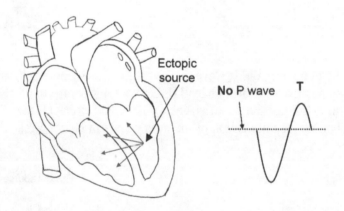

ECG Waves, Contraction and the Cardiac Action Potential

Depolarization of the atria (P-wave) occurs as each atrial cell goes through **phase 0** of the action potential. The **repolarization** of the atria (**phase 3**) occurs during the QRS complex (ventricle depolarization), and so it is not visible.

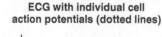

ECG with individual cell action potentials (dotted lines)

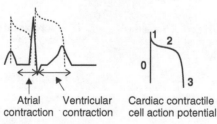

Atrial contraction   Ventricular contraction   Cardiac contractile cell action potential

Depolarization of the ventricles (R-wave) occurs as each ventricular cell goes through **phase 0**. **Repolarization** of the ventricle is associated with **phase 3** of the action potential, and produces the T-wave. **Contraction** begins at the onset of the plateau phase (when Ca^{++} enters the cell), and ends shortly after repolarization is complete.

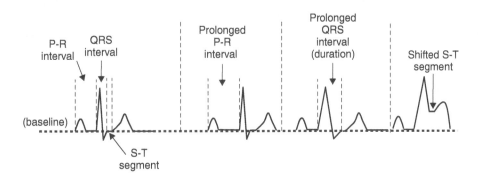

The **P-R interval** (from the beginning of the P-wave to the beginning of the Q- or R-wave) is most closely associated with transit of the depolarization wave through the **AV node**. AV node disease can result in a *slower* transit time for the depolarization wave, and a *longer* P-R interval. The **QRS interval** (from the beginning of the Q- or R-wave to the end of the S-wave) is the time during which the depolarization is spreading throughout the ventricles. A prolonged QRS interval may occur with enlargement (hypertrophy) of the ventricles, or with damage to the conduction fibers (bundle branch block). The **S-T segment** (from the end of the S-wave to the beginning of the T-wave) is a small time span that normally occurs between the spread of the *depolarization* wave through the ventricle and the spread of the *repolarization* wave through the ventricle. The S-T segment should be on the baseline. When a **myocardial infarct** (heart attack) occurs, a portion of the ventricle is deprived of blood flow, and becomes damaged. The depolarization wave is therefore slowed as it travels around the damaged portion, and the R-wave runs into the T-wave, preventing the S-T segment from returning to baseline. Thus, the S-T segment is typically "**shifted off**" the baseline (see above) after a heart attack.

Diseased AV Node- AV Node Blocks

AV node disease results in conduction problems between the atria and ventricles. A **first-degree AV node block** results in a longer than normal transit time of the depolarization through the AV node, which produces a <u>longer</u> than normal P-R interval. Since every P-wave is followed by a QRS, the atria and ventricles still contract at the <u>same rate</u>.

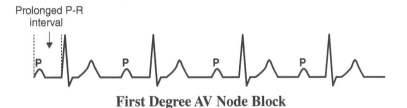

First Degree AV Node Block

A **second-degree AV node block** results when some depolarizations are successfully conducted to the ventricle while other depolarizations do not make it through the diseased AV node. Therefore, the ECG tracing shows some P-waves <u>not followed</u> by QRS- and T-waves, and the ventricular depolarization rate is *slower* (fewer beats per minute) than the atrial rate.

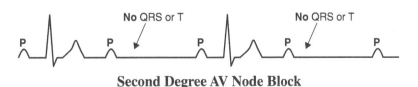

Second Degree AV Node Block

In a **third-degree AV node block**, the AV node does not conduct **ANY** atrial depolarizations to the ventricle. The atria continue to contract at the SA node rate- producing regularly spaced P-waves. However, the ventricles contract at the Purkinje fiber rate (20–40 beats per minute), and these depolarizations are completely unrelated to the SA/AV nodes. Thus, the P-waves and QRS-waves have no consistent association. In other words, the **P-R interval is highly variable**. P-waves also occur without QRS complexes because the atrial rate (SA node rate) is much faster than the ventricular rate. Often the QRS wave is also somewhat altered. Since more P-waves occur than R-waves, the ventricular rate is clearly <u>slower</u> than the atrial rate.

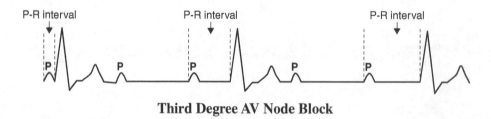

Third Degree AV Node Block

Arrhythmias Originating in the Atria

On occasion, an ectopic source in the atria may beat the SA node to threshold (<u>once</u>), causing the atria to experience a Premature Atrial Contraction (**PAC**). During a PAC, the P-wave may be slightly altered; however the QRS- and T-waves remain normal since depolarization of the ventricle follows the normal pathway.

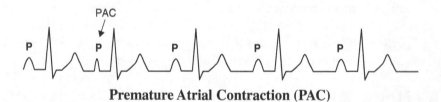

Premature Atrial Contraction (PAC)

If the ectopic source in the atrium depolarizes <u>repeatedly</u> at a rate *faster* than the SA node rate, the new area takes over as pacemaker; this is then called an **ectopic pacemaker**. If the AV node refractory period can accommodate the atrial rate, each atrial depolarization (P-wave) is followed by a ventricular depolarization and repolarization (QRS- and T-waves), and the arrhythmia is known as **atrial tachycardia**. During atrial tachycardia, because the heart rate is so rapid the ventricles do not have adequate time to fill. EDV, SV, and cardiac output (CO) therefore, all decrease, and mean arterial pressure (MAP) may also drop below normal during atrial tachycardia. Since the ventricle of the heart receives blood flow during diastole, with a short diastolic period the muscle of the heart may not receive adequate blood flow which can lead to ventricular muscle damage.

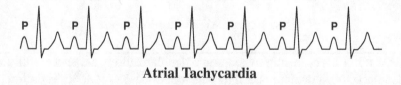

Atrial Tachycardia

If the depolarization rate of the ectopic source in the atria is *faster* than the AV node refractory period can accommodate, some depolarizations will arrive at the AV node when it is still in refraction from the previous depolarization. Therefore, some of the depolarizations will not be conducted to the ventricle (as they arrived at the AV node while it was still in refraction). Thus, the ECG tracing will display some P-waves are *not followed* by QRS- and T-waves. This arrhythmia is known as **Atrial Flutter**. Atrial flutter also produces extremely fast ventricular rates and compromises filling time, resulting in a low CO and MAP. Ventricular muscle blood flow is also reduced.

Atrial Flutter

Sometimes atrial depolarizations become out of "sync," and therefore randomly circle around and around the atria- no longer depolarizing and contracting the atria as a unit. This very common arrhythmia is known as **Atrial Fibrillation**. During atrial fibrillation, the ECG shows a baseline that is very irregular, and the QRS- and T-waves appear erratically- depending on when the atrial waves hit the AV node (if the AV node is out of refraction). Although common, atrial fibrillation is <u>not</u> typically fatal as the ventricle is able to fill ~80% even without the atrial contraction (due to venous return). However, atrial fibrillation is a serious medical condition that should be treated. It is more common in elderly patients, and results in an increased risk of stroke due to an increased likelihood of clot formation in the non-contracting atria. Atrial fibrillation produces extremely fast ventricular rates that compromise filling time, and result in low CO and MAP. Ventricular muscle blood flow is also reduced.

Atrial Fibrillation

Arrhythmias Originating in the Ventricle

A **Premature Ventricular Contraction (PVC)** results from an ectopic source in the ventricle reaching threshold (<u>once</u>) *before* the SA node induced depolarization reaches the ventricle. The abnormal ventricle depolarization produces an **abnormal** (often inverted) **R-wave**, followed by an **abnormal T-wave**.

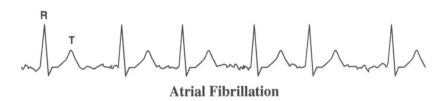

Premature ventricular contraction (PVC)

Note, that ventricular ectopic depolarizations are not conducted backward to the atria because reverse conduction is extremely slow in the AV node. Therefore a **compensatory pause** typically follows a PVC. This pause occurs because the SA node reached threshold sometime *during* the PVC, but the depolarization arrived at the ventricle when it was still in refraction from the ectopic PVC. Note, the normal P-wave is not typically visible due to the large abnormal depolarization (R-wave) and repolarization (T-wave) that occurs during the PVC. However, the P-wave that *would have* been produced by the atrial depolarization, and the QRS that *would have* followed the P-wave (had the ventricle been out of refraction) are shown in *dotted lines* above. Thus, the **compensatory pause** occurs because of a **"skipped beat"** (due to the ventricle refractory period). The contraction strength of a PVC is often weak, due to less than normal filling time and a small EDV. However, the subsequent contraction is often extra strong due to increased filling time and a large EDV.

Ventricular Tachycardia results when an ectopic source in the ventricle depolarizes <u>repeatedly</u> at a rate *faster* than the SA node, and this new area takes over as pacemaker. No P-waves are visible because the strong depolarization/repolarization (R and T) waves override the smaller P-waves, although the atria rhythm is most likely still controlled by the SA node. Owing to the rapid ventricular contraction rate, ventricle filling time is reduced, EDV becomes smaller, SV is reduced, and CO and MAP fall. The rapid contraction rate of the ventricle also interferes with blood flow into the ventricular muscle.

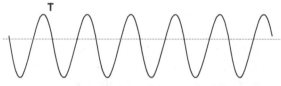

Ventricular Tachycardia

Ventricular tachycardia is a serious condition that can progress to **Ventricular Fibrillation** (see below). During ventricular fibrillation depolarizations become out of "sync," and therefore randomly circle around the ventricles- no longer depolarizing and contracting the ventricles as a unit. The baseline becomes *irregular* and no distinct ECG waves are visible.

Ventricular Fibrillation

Ventricular fibrillation is an extremely <u>serious</u> condition because the ventricles are not contracting as a unit. Therefore, pumping ability is near zero, and CO and MAP dramatically fall to near zero. With inadequate blood pressure, vital organs such as the brain are deprived of blood flow, and **death** results in as little as 2 minutes. **Defibrillation** can be attempted by passing a strong electric current across the chest over the heart. The goal of defibrillation is to force all of the heart muscle cells to depolarize at once, with the hope that the next ventricular depolarizations will occur as a unit.

Experiment: Performing an ECG
Setup

1. One member of your group should perform each of the following functions:
   a. **Director:** Reads the directions.
   b. **Controller:** Runs the computer.
   c. **Subject:** person from whom data are being recorded.
2. Log in to the computer at your station.
3. Turn the MP36 on using the switch at the back of the unit.
4. Double click on the BIOPAC BSL Student Lab 4.1 icon on the desktop.
5. Click on the PRO Lessons tab, select ECG and then click OK.

Electrode Connections

1. Using an alcohol swab, scrub the skin surface at each site (see below) and wipe dry with a paper towel before placing the electrodes.
   • <u>Electrodes:</u>
     • Place one electrode on the inner surface of each leg, just above ankle
     • Place another on the ventral surface of both arms *i.e.*, on the palm side, just above wrist

- Leads:
  - The pinch connectors work like small clothes pins, but must be placed with the metal clip portion down.
  - Attach the black lead to the electrode on the left arm, the red lead to the electrode on the left leg, the white lead to the electrode on the right arm and the green (ground) lead to the electrode on the right leg.
  - **DO NOT CONNECT THE BROWN LEAD!!!!**
  - Clip the electrode cable clip (where the cable meets the five individual colored wires) to a convenient location (such as subject's clothes) to relieve cable strain.
2. Before recording, note the following:
   - Apply the electrodes at least 3 minutes before recording.
   - Nothing should rub or push against the electrodes, including the subject's clothing.
   - Subject should remain still during all recording segments, with his/her hands on their lap, not on the lab bench.
   - Ensure that the electrodes are adhered securely to the skin. If they are coming loose, attach a new electrode to that site.

Data Collection

1. With the subject sitting comfortably and completely still, click on Start.
2. The program will record data for 30 seconds and will then automatically stop.

   Note: The recording is NOT set up to append, so anytime you stop and start, the data will record over the old data and the old data will be erased.
3. Your data should resemble the picture shown in **Figure 7.1**. If it does not,
   a. contact your instructor immediately, or
   b. if errors are known to have occurred during recording, click on Start and then select Yes to Overwrite the existing data

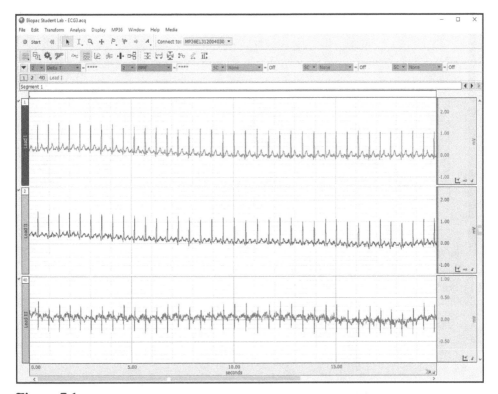

Figure 7.1

4. Save your data (File>Save as) in your section's folder within the Student Data folder on the desktop. Name your file ECG_resting.
5. Remove the leads from the electrodes. **DO NOT REMOVE THE ELECTRODES!!**
6. Run up and down 1 flight of stairs two times or do jumping jacks for 2 mins.
7. Immediately reattach the leads and repeat steps 1-3.
8. Save your data (File>Save as) in your section's folder within the Student Data folder on the desktop. Name your file ECG_after exercise.

Analysis of Data

1. To convert the graphs to the same scale, select Compare Waveforms under the Display menu.
2. Make channels 1 and 40 invisible (See Exercise 2).
3. To magnify your ECG waves:
   a. Click on the Zoom icon (magnifying glass) at the top of the graph window
   b. Drag a box over about 3–4 ECG waves as shown in **Figure 7.2**. If you over zoom, you can always "undo" your zoom by choosing Zoom Back under the Display menu.
   c. The waves should expand and fill the graph window as illustrated in **Figure 7.3**. If any waves appear to be larger than the screen, choose autoscale your data.

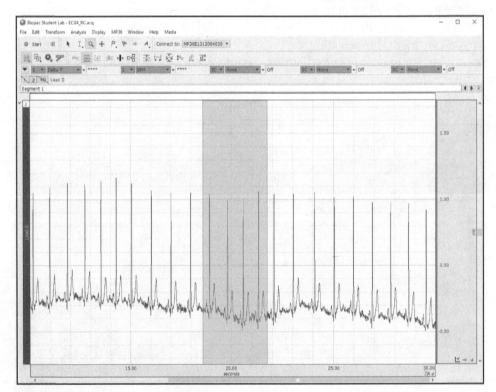

Figure 7.2

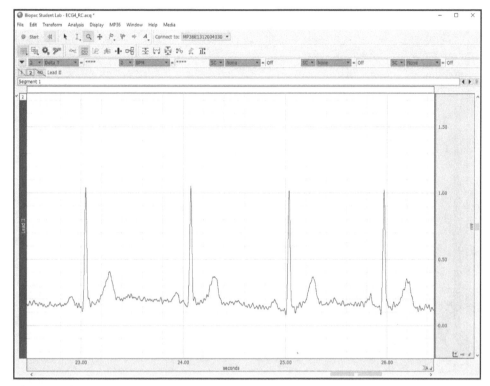

Figure 7.3

I. Measuring the R-R interval (normal = 0.6-1.2 seconds) and Heart Rate (normal = 60-100 beats per minute)

1. Click on the second icon from the left under the Start menu and then on journal in the drop down menu that appears (See Exercise 1, **Figure 1.5**).
2. Click in the journal and type R-R interval/Heart Rate and hit Enter.
3. Click on the I-icon to drag and highlight from the middle of one R-wave to the middle of the next R-wave (See **Figure 7.4**).
4. Press Ctrl-M on your keyboard to record this data in your journal. The R-R interval measurement (duration of one complete cardiac cycle) will be the first of the two numbers recorded and heart rate, the second number.
5. Repeat steps 3 and 4 on a second wave.
6. Average the two heart rates (bpm) and enter the value on your data sheet at the end of this exercise.

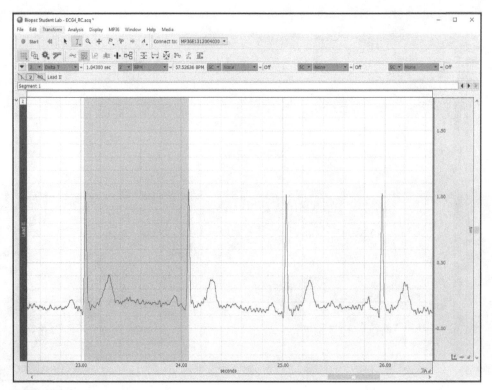

Figure 7.4

II. Measuring the P-R interval (Normal duration = 0.12-0.20 seconds)

1. Click in the journal beneath the already recorded data and type P-R interval and then hit Enter.
2. Adjust the second set of measuring settings at the top of the graph window. To do so, click on the bpm and select none from the drop down list. Your settings should reflect those in presented in **Figure 7.5**.
3. Click on the I-icon to drag and highlight from the beginning of the P-wave to the beginning of the Q- or R-wave (See **Figure 7.6**).
4. Press Ctrl-M on your keyboard to record this data in your journal. Be sure to record this measurement on the data sheet at the end of this exercise.

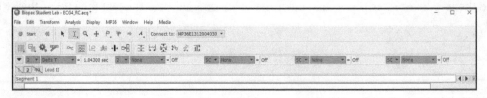

Figure 7.5

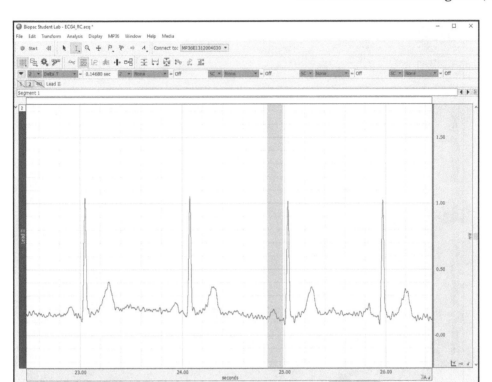

Figure 7.6

III. Measuring QRS interval (normal = 0.10 seconds)

1. Click in the journal beneath the already recorded data and type QRS interval and then hit Enter.
2. Click on the I-icon to drag and highlight from the beginning to the end of one QRS complex (See **Figure 7.7**).
3. Press Ctrl-M on your keyboard to record this data in your journal. Be sure to record this measurement on the data sheet at the end of this exercise.
4. Save your data (File>Save).

IV. Measure R-R interval, P-R interval and QRS interval for the ECG performed after exercising

1. Repeat the steps in I, II and III above.
2. Turn in the data sheet at the end of this exercise to your instructor at the end of your lab session.
3. Close the MP36 graph window and turn the Biopac off via the switch in back of the MP36 unit.

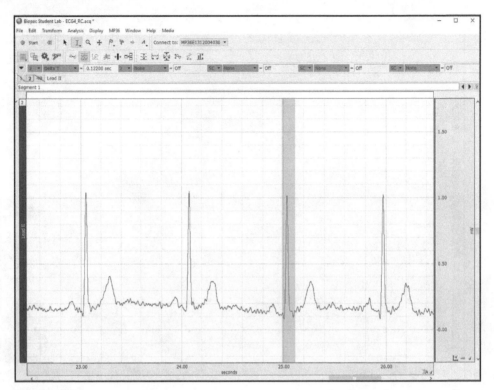

Figure 7.7

Electrocardiogram Data Sheet

Group Number: _____   Name: _____

	At Rest	**After Exercise**
Average R-R interval	seconds	seconds
Average heart rate	bpm	bpm
P-R interval	seconds	seconds
QRS duration	seconds	seconds

a. What is occurring within the heart that the ECG/EKG is actually measuring?

b. Please describe what is happening physiologically during each wave (P, QRS, and T).

c. What observable changes are noted in the waves of the EKGs recorded at rest and after exercise?

d. How does the time for the R-R, P-R and QRS intervals after exercise compare to those recorded at rest? Why do you think this occurs?

Properties of Cardiac Muscle (Turtle Heart)

Study Questions

1. Describe the **pulmonary circulation** and the **systemic circulation**.
2. Why doesn't atrial failure result in death?
3. What are the two parts of the cardiac cycle? Which lasts longer?
4. What part of the cardiac action potential prolongs the action potential and the refractory period? Why is this important? How did we demonstrate this long refractory period in our turtle heart lab?
5. What prevents summation or tetany from occurring in the heart? Why is this important in heart function?
6. Why doesn't the heart experience <u>recruitment</u>? (What characteristic of heart cells prevents recruitment?)
7. How is the excitation–contraction mechanism of the cardiac cell similar to that of skeletal muscle? How is it different from skeletal muscle?
8. What is cardiac output (CO)? How is it calculated?
9. How do the pacemaker cells of the heart get to threshold? What type of channel is involved? How is this channel different from other channels we have studied?
10. Which nervous system speeds the heart rate? Which nervous system lowers heart rate?
11. How can the ventricle contract with more force, <u>without stretch</u>?
    a. How is this different from skeletal muscle?
    b. What is the name for this change?
    c. Which autonomic nervous system and neurotransmitter increases contraction strength by this means?
    d. How does this increase in force of contraction affect stroke volume (SV)? CO?
12. What is end diastolic volume (EDV)? How does EDV affect SV and CO? Explain.
13. If the heart cannot experience summation/tetany or recruitment, by which <u>two</u> mechanisms does the heart increase its strength of contraction?
14. What effect does stimulating the vagus nerve have on the heart? Which branch of the autonomic nervous system is involved?
    a. Name the chemical neurotransmitter released.
    b. What is the receptor?
    c. Describe the effect on the K^+ channels in the pacemaker cell.
    d. Describe the effect on the "funny" Na^+ channels in the pacemaker cells.
    e. What happens to the resting or baseline membrane potential?
    f. How does this slow/stop the heart?
    g. What happens to the acetylcholine after it is released?
    h. After a period of no heart beat, why might the next contraction be stronger than usual?
15. What is the Starling law of the heart? How was this illustrated in our turtle demonstration? How does a normal heart stretch?
    a. What two things happen within the heart cell that causes the Starling effect?
    b. Note how the baseline of the graph shifts up as the heart is stretched, much like the baseline shifted up in our skeletal muscle experiment (See Exercise 6).
16. During what part(s) of the heart turtle heart demonstration would we expect to have an increase in EDV?

17. What is an ectopic beat? Why is an ectopic contraction usually <u>weaker</u> than a normal contraction?
18. Why is an ectopic beat followed by a "skipped beat"?
    a. What is different about the strength of the next ventricular contraction immediately following the skipped beat? Explain.
19. How did we record the depolarization wave in the ventricle (R-wave)? Where does <u>ventricular contraction</u> occur in relationship to this wave (before, during, and after)?
20. Which property of the heart was illustrated by each of the following?
    a. Stretch
    b. Stimulating the ventricle during a contraction
    c. Stimulating the ventricle with greater and greater frequencies.
21. Which property of the heart muscle was NOT demonstrated in the turtle heart lab?

Applied Questions (Answers in back)

1. Well-trained marathon runners have slower resting heart rates than untrained individuals, yet their CO is the same as an untrained individual. Predict the EDV in these athletes. Predict the SV. Explain.
2. Caffeine causes extra Ca^{++} to be released from the sarcoplasmic reticulum (SR) of the cardiac muscle cells in response to an action potential. Predict the effect of caffeine on the contraction strength of the ventricle. Predict the effect on SV and CO. Explain.

Introduction

The purpose of the heart is to act as a pump, creating a pressure difference in the circulatory system. Blood flows from <u>higher</u> to <u>lower</u> pressure. The heart consists of two atria and two ventricles, and the atria contract prior to the ventricles. Our heart is composed of two sides for two different circulations. The **pulmonary circuit** (lungs) consists of blood moving from the right ventricle, through the lungs, and back to the left atrium. The **systemic circuit** consists of blood moving from the left ventricle, through the body tissues, and ends in the right atrium. The ventricles fill 80% full with blood that flows passively from the veins and through the atria. The atrial contractions therefore contribute about 20% of the blood in the ventricle. Atrial contractions are <u>not</u> <u>necessary for life at rest</u>, but are important during exercise.

The cardiac cycle consists of two phases. **Systole** occurs when the ventricles contract and eject blood into the arteries (about 1/3 the time of a single cardiac cycle). **Diastole** occurs when the ventricles relax and fill with blood (about 2/3 the time of a single cardiac cycle). **End-diastolic volume (EDV)** is the amount of blood in the ventricle at the end of filling (prior to ventricular contraction). A larger EDV results in more stretch of the heart muscle. **Stroke volume (SV)** is the amount of blood ejected out of the heart during each ventricular contraction. **End-systolic volume (ESV)** is the amount of blood that remains in the ventricle at the completion of the systolic phase of the cardiac cycle. **Cardiac output (CO)** is the amount of blood pumped by each ventricle per minute. CO is related to both SV and heart rate (HR) such that CO = SV x HR.

A depolarization (excitation phase) always precedes a heart contraction (mechanical phase). The depolarization of the heart can be measured from the body's surface using an electrocardiogram (ECG or EKG). The depolarization of the ventricle is recorded as the QRS complex or R-wave, which occurs right before ventricular contraction. Cardiac muscle contains **gap junctions**, which electrically connects all cardiac muscle cells together. Thus, unlike skeletal muscle, an action potential generated in one cardiac muscle cell will quickly spread to all heart cells, causing all of the cells to contract at about the same time. A special intrinsic conducting system helps coordinate the timing of contractions between various areas of the heart. Therefore, the cells of both atria contract together, followed by contraction of the cells of both ventricles. Because all cells participate in each contraction, **the heart cannot undergo recruitment.**

Unlike skeletal muscle, the cardiac muscle action potential contains a long **plateau phase**, which prolongs the duration of the action potential, causing a long refractory period. This plateau phase is caused by the fact that Ca^{++} channels are open (allowing influx of positively charged ions into the cell) at the same time as K^+ channels are open (allowing efflux of positive ions). Note that the long refractory period lasts almost as long as the mechanical contraction, and therefore the heart cannot be stimulated to contract again until the refractory

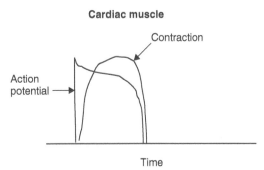

Skeletal Muscle Summation

Action potentials

Contractions

Time

In a skeletal muscle, summation is possible because the action potential (and refractory period) are short.

Cardiac muscle

Contraction

Action potential →

Time

In a cardiac muscle, summation is NOT possible because the action potential (and refractory period) are prolonged.

period and contraction/relaxation is over. Thus, **the heart cannot undergo summation and/or tetany**. This ensures that the heart can relax and fill with blood between contractions.

Excitation-contraction coupling is essentially the same as that for skeletal muscle (review EMG and skeletal muscle exercises). The only difference is that heart muscle can undergo a calcium-induced-calcium-release mechanism where, during the plateau phase of the action potential, extracellular calcium enters the cell and triggers the release of calcium from the SR inside the cell.

Pacemaker cells control heart rate. The cardiac cycle is initiated by action potentials produced from the pacemaker cells of the sino-atrial (SA) node. Importantly, no outside stimulus is required for the pace maker cells to get to threshold. Pacemaker cells depolarize to threshold on their own (**the pacemaker potential**) through the opening of "funny" Na^+ channels (fNa^+) that open when the cell repolarizes. Na^+ (and later Ca^{2+}) enter the cell and brings the membrane to threshold. Although the SA node reaches threshold without any outside influence, the autonomic nervous system can speed or slow the heart rate by altering the pacemaker potential of the SA node. The **parasympathetic nervous system** releases acetylcholine, which binds to M2 muscarinic receptors on the SA node, resulting in extra K^+ channels opening during baseline and fewer fNa^+ and Ca^{2+} channels open during the pacemaker potential. Extra open K^+ channels results in a hyperpolarization in the SA node, increasing the time it takes the membrane to depolarize to threshold and slowing heart rate. Fewer open fNa^+ and Ca^{2+} channels cause an increase in the time it takes for the cell to reach threshold, decreasing the slope of the pacemaker potential and slowing heart rate. The **sympathetic nervous system** releases norepinephrine that binds to β-1 adrenergic receptors on the SA node causing more fNa^+ and Ca^{2+} channels to open during the pacemaker potential. Na^+ and Ca^{2+} enter the cell quickly during the pacemaker potential, bringing the cell to threshold faster (steeper slope of the pacemaker potential) and speeding up heart rate.

The sympathetic nervous system, through the action of norepinephrine, also increases atria and ventricular contraction by increasing contractility. **Contractility** is the ability of heart muscle cells to increase contraction strength by increasing available Ca^{2+}. Norepinephrine binds to β-1 adrenergic receptors on the contractile cells of the heart, causing Ca^{2+} channels to open, bringing in more extracellular Ca^{2+} into the cell. This extra Ca^{2+} causes an increase in contraction strength by increasing the number of crossbridges, resulting in an increased SV and in increase in CO. Please note, that stretch (or EDV) is not altered by contractility.

Similar to skeletal muscle, there is a length-tension relationship in cardiac muscle, such that the force generated by the heart muscle during systole is affected by muscle length. However, unlike skeletal muscle, cardiac muscle in a healthy heart is always <u>shorter</u> than optimal length. Thus, when more blood enters the heart (increasing EDV), cardiac muscle is stretched closer to optimal length, allowing more blood to be ejected from the heart. This is **Starling's Law** of the heart – **the heart pumps out whatever blood comes in**. Increased EDV increases contraction strength by 1) stretching the cardiac sarcomeres to improve overlap of actin and myosin and 2) increasing the <u>affinity</u> of troponin for Ca^{2+}.

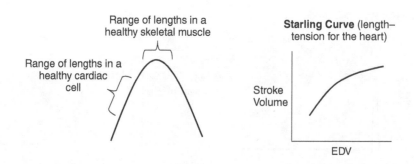

Turtle Heart Demonstration

Experiment 1: The Cardiac Cycle

In this demonstration, we will record the cardiac cycle, including systole (ejection) and diastole (filling) (**Figure 8.1**). **Ventricular systole** begins right after atrial contraction and ends at the peak of the contraction wave. **Ventricular diastole** begins at the peak of the contraction, extends through relaxation <u>and</u> atrial contraction, and ends just as the ventricle begins to contract again (systole). Notice how the ECG R-wave occurs right before the onset of ventricular systole.

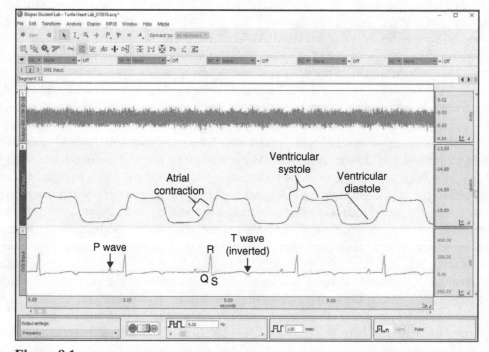

Figure 8.1

Experiment 2: Effect of Stimulating the Right Vagus Nerve

In this experiment, we will apply a high stimulus voltage to the right vagus nerve, creating a high frequency of action potentials. The right vagus nerve is part of the parasympathetic nervous system that innervates the SA node of the heart. When the vagus is stimulated, acetylcholine is released and binds to the muscarinic receptor, hyperpolarizing the cells and slowing the pacemaker potential as described above. Threshold is not reached during the time that the stimulus is applied and the heart does not contract (**Figure 8.2**). Once the stimulus is removed, acetylcholinesterase must break down acetylcholine before contractions resume. Note, that the first contraction after the heart resumes beating is often stronger than usual due to increased blood entering the ventricle when the heart was not beating- which results in an increased the stretch of the heart, increasing the force of contraction (Starling's law).

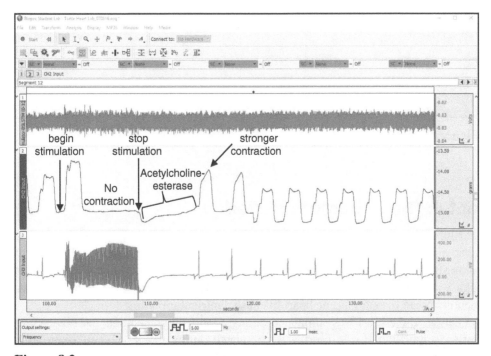

Figure 8.2

Experiment 3: Starling's Law - Effect of Muscle Length on Contraction Force

In this demonstration, the heart is artificially stretched by pulling on a hook in the ventricle and observing the resulting change in contraction strength (height of contraction wave, **Figure 8.3**). This artificial stretch simulates an increased EDV, which results in a stronger contraction (Starling's law).

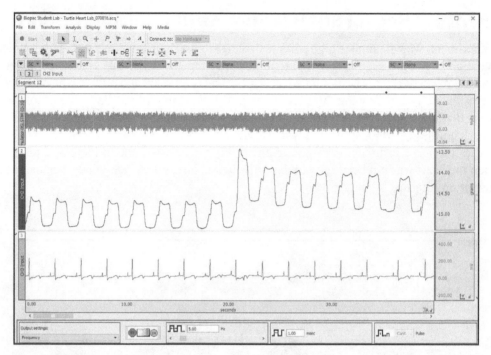

Figure 8.3

Experiment 4: The Refractory Period

In this experiment, we stimulate the heart at various times during the cardiac cycle to demonstrate the prolonged cardiac action potential (mediated by the Ca^{++}/K^+ plateau). In **Figure 8.4**, lines have been drawn from the stimulus signal on Channel 1 to the heart wave on Channel 2 to indicate when during the cardiac cycle the stimulus was applied. The small x's represent normal SA node mediated depolarizations to threshold, which subsequently spread throughout the heart to cause ventricular contraction.

During stimulation A, the heart was stimulated during contraction/relaxation and does not result in another contraction. During stimulation B, the heart was stimulated during the last 1/3 of diastole, causing another contraction to occur (contraction C). This contraction is called an extra-systole, premature ventricular contraction (PVC) or an ectopic beat. An ectopic beat is any contraction that occurs following a depolarization that originates from outside the SA node. The source or origin of the depolarization (outside of the SA node) is called an **ectopic source** or **ectopic pacemaker**. Immediately after contraction C, there is a longer period of time before the next beat called a **compensatory pause** (section D). This is caused because the extra systole has a refractory period, just as a normal contraction does. The SA node reaches threshold as usual, but the depolarization arrives at the ventricle when it is still in refraction from the extra systole. The ventricle will not depolarize or contract, resulting in a skipped heart beat. The ventricle must "wait" for the next SA node induced depolarization to arrive. This pause is the compensatory pause (**Figures 8.4, 8.5**). The contraction after the compensatory pause is a stronger contraction (Contraction E) because there is more time for blood to enter the ventricle, stretching the heart closer to optimal length.

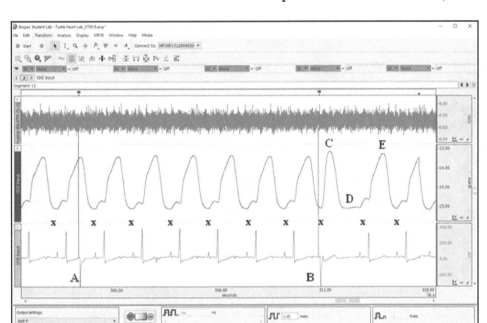

Figure 8.4

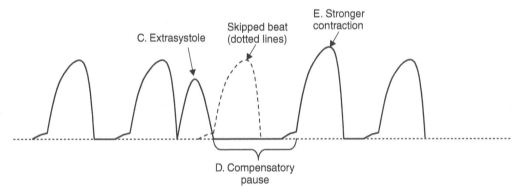

Figure 8.5

Experiment 5: Effect of Stimulus Frequency on Contraction Force

In this experiment, we attempt in the heart to mimic the response of skeletal muscle to an increasing frequency of stimulations, *i.e.*, summation and tetany. Summation and tetany cannot occur due to the long action potential refractory period (as illustrated above). We will increase the frequency of stimulation and find that the ventricle **fibrillates**, or contracts in an asynchronous manner, but the contraction strength does not increase (**Figure 8.6**). This fibrillation, which is shown by erratic depolarizations and contraction waves, will continue for a period of time after stimulation is stopped.

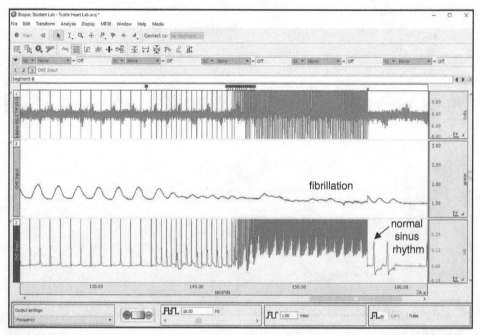

Figure 8.6

Experiment 6: Autogenic Property of the Heart

The heart is <u>autogenic</u> because no outside nervous system stimulation or factor is required to initiate the production of a pacemaker potential from the SA node. The funny Na^+ (fNa^+) channels repeatedly open at repolarization, starting another depolarization to threshold, resulting in another action potential and contraction (**Figure 8.7**). In this demonstration, the heart will be removed from the turtle and will continue to beat outside of the body due to the fNa^+ channels.

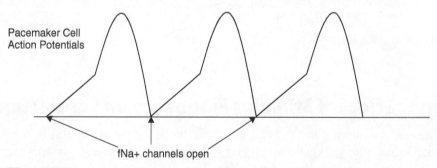

Figure 8.7

Properties of Cardiac Muscle Data Sheet

Group Number: _____   Name: _____

Experiment 1: Cardiac Cycle

Describe the cardiac cycle in relation to the EKG waves.

Experiment 2: Parasympathetic Stimulation

What happened to the heart while the vagus nerve was electrically stimulated? What about after the stimulus was taken away?

Experiment 3: Starling's Law

What happened to the heart while it was being stretched? Please explain this effect physiologically.

Experiment 4: Refractory Period

At which times were we able to stimulate a new contraction during the cardiac cycle? When couldn't we? Why?

Experiment 5: Stimulus Frequency

What happened to the heart when it was stimulated with high electrical frequencies? Did it undergo treppe, summation, and tetany? Why or why not?

Experiment 6: Autogenic Property

What happened to the heart once it was taken out of the turtle's body? Please explain this effect physiologically.

Measurement of Arterial Blood Pressure

Study Questions

1. How is blood pressure measured?
2. What is laminar blood flow? How is it related to the measurement of blood pressure?
3. Why no sound is heard when the sphygmomanometer is **not** inflated?
4. What causes the sounds of Korotkoff?
5. Relate the sounds of Korotkoff to
   a. Systolic arterial blood pressure
   b. Diastolic arterial blood pressure.
6. What is hypertension?
7. What is pulse pressure?
8. Why is mean arterial pressure closer to diastolic pressure than to systolic pressure? How is it calculated?

Introduction

Arterial blood pressure changes over time. **Systolic pressure** occurs during ventricular systole (*young adult male: 120 mm Hg; young adult female: 90–110 mm Hg*). **Diastolic pressure** occurs during ventricular diastole (*young adult male: 80 mm Hg; young adult female: 60–80 mm Hg*). Blood pressure is reported as systolic pressure over diastolic pressure (*i.e.*, 120/80). Individuals with hypertension, or high blood pressure, typically have high systolic ($\geq$ 140 mm Hg) and diastolic ($\geq$ 90 mm Hg) values.

Systolic and diastolic pressure is used to calculate other values that can be important in diagnosing hypertension or other cardiovascular disorders. **Pulse pressure** is systolic pressure – diastolic pressure (*i.e.*, 120 – 80 = 40 mm Hg). Pulse pressure is used to calculate **mean arterial pressure (MAP)**, which is the average arterial pressure over time. Because the heart spends more time in diastole than systole, MAP is closer to diastolic pressure:

$$\text{Diastolic pressure} + 1/3 \text{ (pulse pressure)} = \text{MAP}$$

$$\text{Example: } 80 + 1/3 (120 - 80) = 93.33 \text{ mm Hg}$$

Blood pressure is measured using a sphygmomanometer and blood pressure cuff. Normally, blood flow in arteries is **laminar**, which is smooth, non-turbulent flow. This means that blood flow is normally silent. To measure blood pressure, a cuff is placed over the brachial artery and inflated to a pressure that is greater than the expected systolic pressure. The high pressure stops blood flow through the brachial artery and no sounds are heard. Cuff pressure is gradually decreased and when cuff pressure = systolic pressure, blood spurts through the vessel with each heart beat. This creates a turbulent (noisy) flow. The first appearance of a sound (first sound of Korotkoff) is systolic pressure. As cuff pressure continues to decrease, the sounds are heard. When cuff pressure = diastolic pressure, the sounds disappear (last (or second) sound of Korotkoff). The sound disappears because blood flow is back to smooth laminar flow. This indicates diastolic pressure.

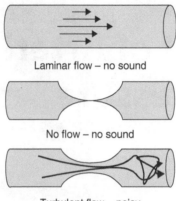

Laminar flow – no sound

No flow – no sound

Turbulent flow – noisy

Subject Setup

1. Remove all air from the blood pressure cuff by turning the release valve all the way in a counter clockwise direction (when looking down on the knob) and rolling up the cuff to press the air out.
2. Wrap the cuff around the subject's right arm, 2 inches above the bend in the elbow. The "artery patch" on the cuff should be in the front of the arm. If not already on the cuff, the pressure gauge can be hooked into the front of the cuff.
3. Place the ear pieces of the stethoscope into your ears, and firmly press the stethoscope diaphragm into the subject's arm over the brachial artery (See **Figure 9.1**). The diaphragm should be below the cuff, but above the elbow bend and toward the middle part of the arm.

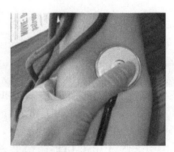

Figure 9.1

IMPORTANT NOTES

- Before you begin the experiment, read through the entire data collection procedure.
- When measuring blood pressure:
    - be sure to hold the diaphragm as still as you can
    - make sure that nothing besides the subjects' arm is touching the diaphragm, including the cuff or cuff tubes
    - ensure that all group members (other than the subject) can see the pressure gauge on the cuff

Data Collection

1. Close the valve on the blood pressure cuff by turning the release valve knob on the bulb completely to the right (clockwise when looking down on the knob). **DO NOT TIGHTEN!!**
2. Once the valve has been closed, squeeze on the bulb several times until the pressure gauge indicates ≈160 mmHg.
3. Turn the valve slightly counterclockwise to open a bit and slowly release air.
4. Listen for Korotkoff sounds through the stethoscope.
   - When the first sound is heard, note the cuff gauge pressure - this is **systolic pressure.**
   - When the sounds disappear, note the gauge pressure where the sound was last heard - this is **diastolic pressure.**
   - Record these values on the data sheet at the end of this exercise.
5. Let all the air out of the cuff.
6. Repeat steps 1-5 for each group member as the subject.
7. Turn in the data sheet at the end of this exercise to your lab instructor.

Blood Pressure Data Sheet

Group Number: _____   Name: _____

Subject	Measurement	Pressure determined by the sound heard through the stethoscope/ sphygmomanometer blips	Pulse Pressure	Mean Arterial Pressure
1	Systolic pressure			
	Diastolic pressure			
2	Systolic pressure			
	Diastolic pressure			
3	Systolic pressure			
	Diastolic pressure			

Why is mean arterial pressure mostly determined by diastolic pressure? Use your knowledge of the cardiac cycle to answer this question.

Regulation of Mean Arterial Pressure (MAP)

Study Questions

1. What is the relationship between total peripheral resistance (TPR) and mean arterial pressure (MAP)?
2. How does a vasoconstriction affect TPR?
   a. How does a vasoconstriction affect MAP?
   b. How does a vasoconstriction affect capillary blood flow?
   c. How does a vasoconstriction affect peripheral pulse pressure (PPP) amplitude?
3. How would an increase in blood viscosity affect MAP?
4. What is the <u>first</u> change in the cardiovascular system that occurs as a result of hemorrhage?
   a. List the sequence of events beginning with hemorrhage and ending with a decrease in MAP.
   b. List the sequence of events initiated by the cardiovascular center, to correct the decrease in MAP. Be sure to include the following:
      1) All the ways the sympathetic nervous system acts to raise blood pressure back to normal.
      2) The ways the parasympathetic nervous system acts to raise blood pressure back to normal.
5. When a person lies down from a standing position, what is the <u>first</u> change in the cardiovascular system?
   a. How does the cardiovascular center (baroreceptor reflex) respond to correct this change?
6. What effect does sympathetic nervous system stimulation have on the peripheral pulse amplitude (PPP)? Why?
7. Why does an increase in TPR cause a small amplitude PPP?
8. List the sequence of events during the Valsalva maneuver that result in a drop in blood pressure.
   a. Explain why the PPP amplitude became small, when the body responded to the Valsalva maneuver?
9. Predict what happens to PPP when standing up from a sitting position. Be sure to determine 1) the change in MAP caused by standing up, and 2) the cardiovascular center's corrective response that alters MAP.

Applied Questions (Answers in Back)

1. Debbie Dry has run a marathon, didn't drink enough water, and has become very dehydrated.
   a. What change would dehydration cause in the cardiovascular system?
   b. Debbie's blood pressure is normal. How can this be so?
   c. What would you predict about Debbie's heart rate? Her TPR? What about her pulse pressure?
2. Artificial respiration is performed by blowing air into the patient's lungs. This is the opposite of a normal, "suction" inhalation. What affect might artificial respiration have on venous return (VR)?
3. Barry Bleed has been in an accident and lost almost 2 pints of blood.
   a. Predict Barry's VR.
   b. Predict Barry's cardiac output (CO).
   c. When lying down, Barry has a normal MAP, even though his CO is low. How can this be so?
4. When standing up, Barry's MAP is below normal.
   a. Predict Barry's heart rate when standing.
   b. Predict Barry's TPR when standing.
   c. Predict Barry's CO when standing.

Introduction

The maintenance of the correct arterial pressure is extremely important in cardiovascular function. Blood in the vascular system flows from regions of higher pressure to regions of lower pressure, and blood flow through the capillaries must be adequate to maintain proper function. The **mean arterial pressure (MAP)** is the average pressure over time in the large arteries, and is the force that is effective for driving blood into the tissue capillaries. MAP is regulated by **cardiac output (CO)** and **total peripheral resistance (TPR)**, such that MAP = CO x TPR.

CO is the amount of blood pumped out of the heart over time. An increase in CO forces more blood into the arteries, and causes an increase in MAP. A decrease in CO results in less blood in the arteries, and causes a lower MAP. CO is determined by **heart rate (HR)** and **stroke volume (SV)**, such that CO = HR x SV. HR, the rate at which the heart beats, is regulated by the parasympathetic and sympathetic nervous system and directly changes CO, such that $\uparrow$ HR = $\uparrow$ CO = $\uparrow$ MAP. SV is the amount of blood ejected from the heart in one contraction and is controlled by **contractility** and **end diastolic volume (EDV)**. Contractility is a change in the contraction strength of the heart due to additional Ca^{2+} available to the sarcomeres. Contractility is controlled by the sympathetic nervous system, such that increased sympathetic activity causes $\uparrow$ Contractility = $\uparrow$ SV = $\uparrow$ CO = $\uparrow$ MAP. EDV is the amount of blood that fills the heart at the end of diastole. Because the heart muscle is shorter than optimal length, as more blood fills the heart ($\uparrow$ EDV), the muscle is stretched closer to optimal, which increases overlap of actin and myosin and produces a stronger contraction ($\uparrow$ EDV = $\uparrow$ SV = $\uparrow$ CO = $\uparrow$ MAP). The relationship between stretch and contraction strength is **Starling's law** of the heart (length-tension relationship). EDV is determined by **venous return (VR)**, the amount of blood going back to the heart from the veins. VR directly changes EDV, such that $\uparrow$ VR = $\uparrow$ EDV = $\uparrow$ SV = $\uparrow$ CO = $\uparrow$ MAP.

VR can be modified in five different ways: 1) skeletal muscle pump, 2) respiratory pump, 3) blood volume, 4) body position, and 5) venomotor tone. The **skeletal muscle pump** occurs when skeletal muscles contract (such as during exercise) and squeeze the veins. As the vessels are squeezed, blood is pushed towards the heart due to one-way valves in the veins that prevent backflow of blood (thereby $\uparrow$ VR). Lack of movement (standing still, for example) causes blood to pool in the leg veins and less blood returns to the heart (thereby $\downarrow$ VR). The **respiratory pump** occurs during inhalation, which creates a negative pressure in the thorax and pulls blood toward the heart, increasing VR. Positive pressure in the thorax would decrease the respiratory pump and decrease VR. An increase in the total **blood volume** increases VR, simply because there is more blood available to go back to the heart. Similarly, a decrease in blood volume decreases VR. **Body position** can affect VR through the effects of gravity. Standing up decreases VR while sitting down increases VR. Finally, **venomotor tone** is the contraction of smooth muscle in the veins that increases blood pressure in the vein, helping to push the blood back to the heart. Increased sympathetic nervous system activation causes an increase in venomotor tone, which increases VR.

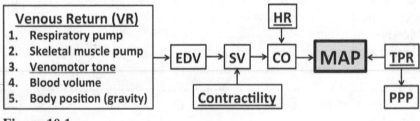

Figure 10.1

TPR is the overall vascular resistance in the systemic circulation, mainly due to arteriole diameter. Vasodilation of the arterioles (increased diameter) decreases TPR and allows blood to leave the arteries at a faster rate, resulting in less blood in the arteries and a <u>lower</u> MAP. Vasoconstriction of the arterioles (decreased diameter) increases TPR and slows the rate that blood leaves the arteries, causing more blood to stay in the arteries and a <u>higher</u> MAP (See **Figure 10.2**).

Vasodilation of the arteriole (↓ TPR)

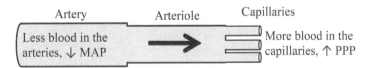

Vasoconstriction of the arteriole (↑ TPR)

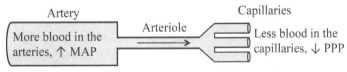

Figure 10.2

Regulation of MAP

MAP is regulated by negative feedback. **Baroreceptors** are the sensors in this system and are located in the carotid sinus and aortic arch. Baroreceptors are stretch receptors that respond to changes in MAP such that an increase in MAP increases the number of action potentials produced by the sensory neuron, and a decrease in MAP decreases the number of action potentials. The integrator in this system is the **cardiovascular center** in the medulla, and the efferent pathway is the efferent nerves of the **autonomic nervous system**. Effectors include the heart, arterioles, and large veins.

The body continually adjusts CO and vascular resistance (arteriolar radius) to maintain our MAP at normal values. Each time a change occurs in the cardiovascular system, the body responds to correct the change in MAP. It is helpful to think of this regulation in two steps:

1. A change occurs that affects MAP (increase or decrease).
2. The cardiovascular center responds to correct the change in MAP.

For example, hemorrhage (blood loss) and dehydration both decrease blood volume. The change (problem) is that there is a ↓ VR. If there is a decrease in VR, then ↓ EDV = ↓ SV = ↓ CO = ↓ MAP. The change in MAP is sensed by the baroreceptors and the cardiovascular center responds by <u>activating</u> the sympathetic nervous system and <u>deactivating</u> the parasympathetic nervous system:

<u>↑ Sympathetic (Four Effects):</u>

1. ↑ HR = ↑ CO = ↑ MAP
2. ↑ TPR (vasoconstriction of arterioles) = ↑ MAP (smaller PPP wave on computer)
3. ↑ venomotor tone = ↑ VR = ↑ EDV = ↑ SV = ↑ CO = ↑ MAP
4. ↑ contractility = ↑ SV = ↑ CO →= ↑ MAP.

<u>↓ Parasympathetic (One Effect):</u>

1. ↑ heart rate = ↑ CO = ↑ MAP.

Our experiment measures **PPP**, which indicates blood flow in the capillaries. Remember that MAP is altered by TPR such that vasoconstriction of the arteriole increases TPR and decreases the amount of blood leaving the artery and entering the capillary. Therefore, in this example, PPP wave amplitude would decrease because less blood is entering the capillaries.

In today's lab, we will perform two manipulations to alter MAP, the Valsalva maneuver and gravity. On the computer, we will <u>not see the original change induced</u>, but will view only some of the body's **corrective responses** to the manipulation. The first experiment will use the Valsalva maneuver to manipulate MAP. The **Valsalva maneuver** is defined as a forced exhalation (blowing) against a closed glottis (closed throat so no air can escape) or against some form of resistance (i.e. blowing up a balloon or playing a trumpet). This maneuver creates a large positive pressure in the chest (thorax), which reduces blood flow to the chest cavity, decreasing VR. The second experiment will use **gravity** to manipulate MAP. We will do this by changing body position (i.e., sitting vs. standing). When standing, blood must travel upward against gravity to return to the heart, decreasing VR.

Setup

1. One member of your group should perform each of the following functions:
    a. **Director:** Reads the directions.
    b. **Controller:** Runs the computer.
    c. **Subject:** person from whom data are being recorded.
2. Log in to the computer at your station.
3. Turn the MP36 on using the switch at the back of the unit.
4. Double click on the BIOPAC BSL Student Lab 4.1 icon on the desktop.
5. Click on the PRO Lessons tab, select PPP - Regulation of MAP and then click OK.
6. One channel, channel 1, which records the PPP wave, should be apparent in the graph window.

Subject Setup

1. Attach the pulse transducer snugly to any finger except for the thumb using the expandable black strap. The wire of the pulse transducer should come off the fingertip (*i.e.*, it should be pointed away from the body). It should feel secure when wiggled, but not too tight.
2. The chair should be positioned so that the subject can stand without having to slide the chair back.
3. The subject should be relaxed, with his/her arm and hand resting on the table, and feet on the circular foot support of the chair or on the floor.
4. The subject must minimize his/her hand and finger movements, which will alter the recording.

Calibration

1. Click on Start in the upper left corner of the graph window. Record the subject's pulse pressure for 15 seconds (watch the time scale on the bottom of the graph window) and then click on Stop. **DO NOT REMOVE THE PULSE TRANSDUCER!!**
2. Autoscale your data if necessary.
3. If your data does not resemble **Figure 10.1**, contact your instructor before proceeding.
4. Save your data (File>Save as) in your section's folder within the Student Data folder on the desktop. Name your file PPP.

IMPORTANT NOTES

- Before you begin the two experiments below, read through the entire data collection procedure for each.
- The subject chosen should have no health disorders that would make this procedure a risk, including any known heart, kidney, or respiratory disorders.
- The subject should stop immediately if dizziness or nausea occurs.

Experiment 1: The Valsalva Maneuver

Procedure

1. Insert a marker event at the end of the 15 - second calibration period by "right clicking" in the bar below the event bar. In the dropdown menu that appears, click on "Insert New Event."
2. Right click on the marker triangle and click Edit event. Highlight and delete any writing in the marker text box, and type 'Valsalva-baseline.'

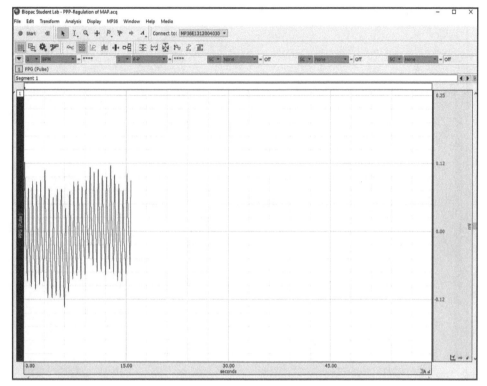

Figure 10.1

3. Click on Start. Obtain approximately 15 seconds of control (baseline) tracing. **DO NOT STOP RECORDING!!**
4. Insert a marker (push the Esc key) just as the subject inhales deeply and tries to blow air out, but does not let any air out because the throat is kept closed.
   - Be sure to create a pressure in the chest by attempting to blow. Just holding your breath will have no effect. Continue this for several seconds or until a response is clearly seen in the waves on the computer
5. Insert a marker (click on the Esc key) just as the subject stops the Valsalva (lets the air out).
6. The subject should then breathe normally for 20 seconds to record the recovery period. **DO NOT STOP RECORDING UNTIL THE RECOVERY PERIOD IS OVER!!**
7. Press Stop.
8. Scroll back and label each of the inserted markers (**Figure 10.2**).
   - The marker after the 'Valsalva-baseline' marker should be labeled 'begin Valsalva'.
   - The next marker, the one inserted when the subject stopped the Valsalva, should be labeled 'end Valsalva'.
9. Save your data (File>Save).

Experiment 2: The Effects of Gravity (Body Position)

Procedure

1. Insert a marker event at the end of the recovery period for experiment 1 by right clicking in the bar below the event bar. In the dropdown menu that appears, click on Insert New Event.
2. Right click on the marker triangle and click Edit event. Highlight and delete any writing in the marker text box, and type 'Sitting' (See Exercise 3, Figure 3.2).

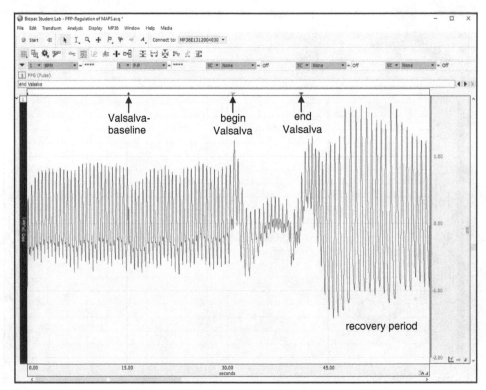

Figure 10.2

3. The subject should be sitting, with his/her feet up on a second chair or on the lab desk. His/her PPP-measured hand should be resting on the lab bench.
4. Click on Start to obtain approximately 20 seconds of control tracing.
5. Insert a marker (click on the Esc key) just as the subject stands up, keeping his/her hand on the lab bench.
   • The subject must stand up with as little overall movement as possible. Try to keep your body and hand very still, by locking your knees and keeping your legs somewhat relaxed.
6. After 20 seconds, insert a marker (click on the Esc key), and the subject should sit down, again (with feet back on the second chair or lab desk) with as little overall movement as possible.
7. The subject should then breathe normally for 20 seconds to record the recovery period. **DO NOT STOP RECORDING UNTIL THE RECOVERY PERIOD IS OVER!!**
8. Press Stop.
9. Scroll back and label each of the inserted markers (**Figure 10.3**).
   • The marker after the 'sitting' marker should be labeled 'standing'.
   • The next marker, the one inserted when the subject sat back down, should be labeled 'sitting again'.
10. Save your data (File>Save).

20 sec sit w/ legs up

20 sec stand up

20 sec sit w/ legs up

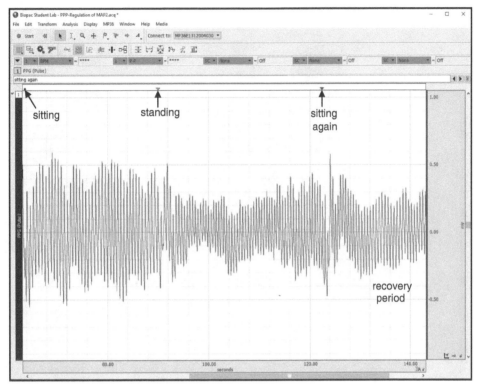

Figure 10.3

Data Analysis: Measuring Heart Rate and PPP Amplitude

A. Valsalva Maneuver

1. Click on the second icon from the left under the Start menu and then on journal in the drop down menu that appears (See Exercise 1, **Figure 1.5**).
2. In the journal window that appears at the bottom of the graph window, type Valsalva and hit Enter.
3. Adjust your time scale to span 3 milliseconds (See Exercise 3, Experiment 2, steps 4a - c).
4. Find your 'Valsalva-baseline' marker using the left and right arrows (◄►) on the right hand of the screen.
5. Measure a baseline wave. To do so, scroll to the right to find a relatively stable section of waves after the Valsalva-baseline marker but before the 'begin Valsalva' marker. Click on the I-icon to drag across single wave, *i.e.*, from the beginning of the wave to the beginning of the next wave (**Figure 10.4**). Be sure to include any space between the waves.
6. Press Ctrl-M on your keyboard. Two values, heart rate (first value) and wave amplitude (second value), will appear in your journal. Record these values on the data sheet at the end of this exercise.
7. Repeat step 5 to measure a wave during the Valsalva maneuver and another during the recovery period (*i.e.*, at least 3–5 seconds after the end Valsalva marker but before the sitting marker that represents the start of Experiment 2).
8. Record these values on the data sheet at the end of this exercise.

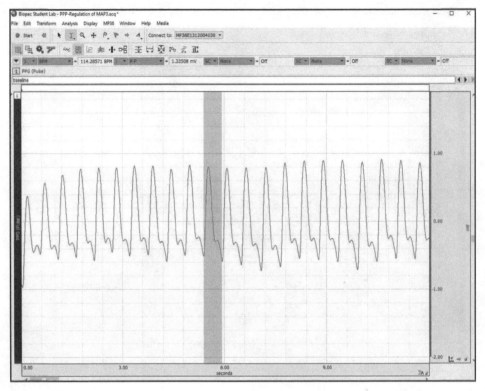

Figure 10.4

B. Effect of Body Position

1. Type Body position (below the Valsalva maneuver data) in the journal window at the bottom of the graph window and hit Enter.
2. Find your 'sitting' marker using the left and right arrows (◄►) on the right hand portion of the screen.
3. Measure, as described in part A, typical waves in stable sections of the data in the 'sitting,' 'standing up,' and 'sitting again' sections of your data.
4. Record these values on the data sheet at the end of this exercise.
5. Save your data (File>Save).
6. Close the MP36 graph window and turn the Biopac off via the switch in back of the MP36 unit.

Regulation of Map Data Sheet

Group Number: _____   Name: _____

Fill in the responses that you measured, and then answer the questions below.

Experiment 1: Valsalva Maneuver

	Heart rate (column 1)	**PPP (column 2)**
Control /baseline		
Response (during)		
Recovery (after)		

1. What effect did the Valsalva maneuver produce on MAP - <u>before</u> any corrective response (increase/ decrease)? Explain.

   a. Which aspect of the autonomic nervous system was activated to correct the "Valsalva-induced" problem? Which one was deactivated?

   b. What *should have happened* to heart rate as your body responded to <u>correct</u> the Valsalva-induced changes (increase/decrease)? Did your results agree?

   c. What *should have happened* to PPP amplitude as your body responded to <u>correct</u> the Valsalva-induced changes (increase/decrease)? Did your results agree?

2. What effect did <u>stopping</u> the Valsalva maneuver have on MAP–<u>before</u> there was any correction (increase/decrease)? Explain.

a. What nervous system was activated to correct the problem? Which one was deactivated?

b. What *should have happened* to heart rate as your body responded to <u>correct</u> the changes induced by <u>stopping</u> the Valsalva maneuver (increase/decrease)? Did your results agree?

c. What should have happened to PPP amplitude as your body responded to <u>correct</u> the changes induced by <u>stopping</u> the Valsalva maneuver (increase/decrease)? Did your results agree?

Experiment 2: The Effects of Gravity (Body Position)

	Heart rate (column 1)	PPP (column 2)
Sitting		
Standing		
Sitting		

3. What effect did standing up from a sitting position have on MAP - <u>before</u> there was any correction (increase/decrease)? Explain.

a. What nervous system was activated to correct the problem? Which one was deactivated?

b. What *should have happened* to heart rate as your body responded to <u>correct</u> the changes induced by standing up (increase/decrease)? Did your results agree?

c. What should have happened to PPP amplitude as your body responded to <u>correct</u> the changes induced by standing up (increase/decrease)? Did your results agree?

4. After you sat down again, what was the effect on MAP - <u>before</u> there was any correction (increase/ decrease)? Explain.

   a. What nervous system was activated to correct this problem? Which one was deactivated?

   b. What should have happened to heart rate as your body responded to <u>correct</u> the changes induced by sitting down (increase/decrease)? Did your results agree?

   c. What should have happened to PPP amplitude as your body responded to <u>correct</u> the changes induced by sitting down (increase/decrease)? Did your results agree?

Blood Types

Study Questions

1. What blood type is the <u>universal donor</u>? The <u>universal recipient</u>?
2. Using your knowledge of the immune system, explain why transfusion with the wrong blood type causes such a severe reaction.
   a. Be sure to include antibodies and antigens in your explanation.
3. Of the following blood types, which transfusions would likely result in a <u>severe reaction</u>?
   a. Type A to B
   b. Type A to AB
   c. Type A to O
   d. Type O to AB
   e. Type O to A
   f. Type O to B
4. What problems occur with the Rh factor and pregnancy?
   a. Why is the first pregnancy not affected by a difference in Rh blood type?
   b. How can the Rh problem be prevented?
   c. Explain the physiology of the immune system as it relates to prevention of the Rh factor.
5. How is blood typed?
6. What blood types are each of the following:
   a. Anti A: clumped; Anti B: clumped; Anti D: clumped?
   b. Anti A: no reaction; Anti B: clumped; Anti D: no reaction?
7. A patient has A^+ blood. Besides A^+, which three additional blood types could safely be transfused into this patient?

Introduction

Blood typing is based on antigens present in the cell membranes of erythrocytes (red blood cells, RBCs). There are two main blood group systems in humans, including the ABO and Rh systems. There are other blood groups, but they are less common.

ABO Blood Group

There are two antigens possible in the ABO system: A and B antigens. Type O refers to the absence of A and B antigens. A and B antigens are closely related to common bacterial or food antigens. Exposure to these food or bacterial antigens activates memory B cell formation and the production of antibodies in the blood against the antigen. If an individual is transfused with the wrong blood type, the antibodies present in the recipient's plasma binds to the antigens on the donated RBCs, initiating a potentially fatal immune response in the recipient. The antibodies in the donated blood are removed prior to transfusion and would not affect the recipient. Thus, an important rule to remember is that the recipient cannot be given any antigens they don't already have. For example, a recipient with AB blood could be given a transfusion of A blood because the A antigen is already present on the recipient's blood.

Blood type	Antigens on RBC	Antibodies in plasma
A	A antigens	Anti-B
B	B antigens	Anti-A
AB	A and B antigens	None
O	None	Anti-A and Anti-B

Rh Blood Group

Rh factor is another possible antigen that is commonly present on RBCs. If the Rh is present on the RBCs, then the blood is Rh^+. For example, A^+ blood has both the A antigen and the Rh antigen on the RBCs. If the Rh antigen is not present, then the blood is Rh^- (i.e. B^- blood has only B antigens on the RBCs). Rh factor is not closely related to other antigens and antibodies against Rh are rarely present in the plasma. However, if an Rh^- person is exposed to Rh^+ blood, they will develop antibodies against Rh. This is especially a problem during pregnancy if the mother is Rh^- and the child is Rh^+. During pregnancy, virtually no mixing of the maternal and fetal blood occurs. However, during birth the mother's blood mixes with the baby's blood, exposing the mother to the Rh antigen. This stimulates formation of memory B cells in the mother and all subsequent pregnancies with a Rh^+ fetus will cause her immune system to produce Rh antibodies. These antibodies cross the placenta and bind to the baby's Rh-antigens, stimulating the babies own immune system to destroy fetal RBCs, which could be fatal to the baby. Unfortunately, once the mother becomes sensitized to Rh (creates Rh antibodies), there aren't any treatments. Therefore, treatment to prevent memory B cell formation in the mother is necessary. To do this, the mother will be given **passive immunity** to Rh within 72 hours of birth for every Rh^+ baby. The mother is given Rh antibodies, which cover the Rh antigens on the fetus RBCs, stopping the stimulation of the mother's immune system and the formation of memory B cells. Rh antibodies must be given during every pregnancy with an Rh^+ fetus.

Blood type	Antigens on RBC	Antibodies in plasma
+	Rh antigens	None
−	None	Anti-Rh (possibly)

Blood Typing

To determine a patient's blood type, antibodies are used to detect the presence of antigens. A patient's blood is drawn and mixed with antibodies to A, B and Rh in separate test tubes. If the antigen is present, then the antibodies will react and cause the blood to clump. For example, if the blood is mixed with antibodies to A and there is no "clumping" reaction, then there are no A antigens on the RBCs. If the blood clumps when mixed with B and Rh antibodies, then there are B and Rh antigens on the patient's RBCs. This individual would be B^+. In this experiment, you will determine three simulated patients' blood types, and then determine which blood transfusions would be safe to give them.

Experiment: Blood–Typing Play the Blood Typing Game

1. Click twice on the **Blood Typing Game** icon on the desktop. If the icon is not present, open the internet browser and type Nobel Prize blood typing game in the search box (the web address is https://www .nobelprize.org/educational/medicine/bloodtypinggame/).
2. When the internet site appears, click on 'Play the Blood Typing Game'. Select Proceed (bottom right) on the next screen to start.
3. Select 'Quick game - same patients,' and then Proceed on the next screen.
4. Select one of the three patients.
5. Click and drag the syringe to the crook of the elbow of the patient's arm and release the button, to get a blood sample.
6. Click/drag the syringe so the needle is placed into the first of the three test tubes, containing antibody A. Release the mouse button to eject some blood. Repeat for the remaining two test tubes containing antibody B and Rh, respectively.
7. Read the results and select the blood type. Remember that clumping in the test tube indicates an antibody-antigen reaction (e.g., if the blood clumps with antibody A, then A antigens are present on the red blood cells, indicating that the individual is blood type A).
8. Click the X to close the 'You're bloody right' window and the transfusion page will appear.
9. Drag a bag of the appropriate type of blood and hang it on the IV stand on the left. Release the button on the mouse to transfuse the blood into the patient. You may need to click the right arrow to see all the possible types available for transfusion.
10. If the patient needs extra blood (the number of bags needed is shown to the left of the transfusion bags), you need to pick another blood type that won't cause a reaction in the patient. Drag the second, or more, bag(s) of blood to the IV stand.
11. When the game indicates you are finished with that patient, choose another patient and repeat.

Blood Typing Data Sheet

Group Number: _____   Name: _____

Patient	Blood type	Possible blood types for transfusion
1		
2		
3		

Why can't a patient who is A⁻ be given B⁺ blood during a transfusion? Answer these questions using your knowledge of antigens and antibodies.

Blood Cells

Study Questions

1. Why is the red blood cell (RBC) important?
2. What is methemoglobin? Why is this important clinically? What might cause it?
3. What is carboxyhemoglobin? Why can this cause death, even when carbon monoxide concentration in the air is *lower* than oxygen levels?
4. What are the two types of anemia?
   a. How does anemia affect the blood's ability to carry oxygen to the tissues?
5. What are the two general categories of white blood cells?
6. What are the important morphological and staining characteristics of each type of blood cell? What is the major function of each?
   a. Red blood cell
   b. Neutrophil
   c. Eosinophil
   d. Basophil
   e. Monocyte
   f. Lymphocyte
   g. Platelet
7. Which cells destroy invading pathogens by phagocytosis? Which is the most phagocytic?
8. What disorders are likely to raise the number of eosinophils in the blood?
9. What disorder is likely to raise the number of neutrophils in the blood?
10. Which cell is involved in anaphylactic shock?
11. What is a mast cell?
12. Which cell types are commonly found in pus?
13. What are platelets? From what cell are they formed? What do they do?
14. What is the difference between nonspecific immunity and specific immunity?
    a. Which cell types are involved in each?
    b. Which is long lasting?
15. What is the difference between B-cells and T-cells?
16. Describe the clonal selection theory.
17. Why is an immune response to a second and subsequent exposure to an antigen greater than the initial immune response? What cell type is responsible for this?
18. How is active immunity different from passive immunity? How is each obtained? Which is longer lasting?
19. How are plasma cells created, and what is their function? What about memory cells?
20. What are the two types of effector T-cells?
21. In what disorder are helper T-cells deficient?
22. What is an autoimmune disease?
    a. What are two possible causes?

23. How is a differential white blood count performed?
24. Why is the total number of white blood cells important?
    a. How is the actual number of each cell type calculated? Why is this important?

Applied Questions (Answers in back)

1. Puppies are commonly vaccinated several times when young because of immunity (maternal antibodies) transferred from the mother to the puppies during pregnancy and through the milk.
   a. Why would the maternal antibodies in the puppy prevent the puppy from producing immunity from the vaccine?
   b. What kind of immunity do puppies get from their mother (active or passive)?
   c. If a puppy has immunity obtained from the mother, why does he need to be vaccinated?
2. If a person is exposed to a specific influenza virus (*i.e.*, H1N1), why isn't he immune to a different influenza virus (H2N2)?
3. One type of heart damage is believed to be caused by the body's immune response to <u>Strep</u> bacteria. How can the immune system's response to a virus cause damage to a patient's own body?
4. Jane Doe has a neutrophil count of 55% (normal = 55–75%). Her total white count is 2000 cells per cubic mm (normal = 5000–10,000). Does she have a normal <u>number</u> of neutrophils?

Introduction

Blood is comprised of plasma, erythrocytes (red blood cells, RBCs), platelets and leukocytes (white blood cells). Plasma is mostly made up of water (92%) with some ions and proteins. Albumin is the most abundant protein in plasma. Plasma makes up the majority of the blood, with approximately 55% of the blood volume due to plasma. The **hematocrit**, or the amount of volume due to red blood cells, can be determined by placing a blood sample into a capillary tube and spinning it at a high speed. The cells will collect at the bottom and the proportion of blood volume that is made up of cells can be determined. Normally, the hematocrit should be approximately 45%.

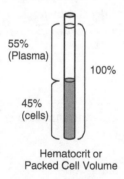

55%
(Plasma)

100%

45%
(cells)

Hematocrit or
Packed Cell Volume

Erythrocytes and Platelets

Hematopoiesis (blood cell production) occurs in the bone marrow. Hematopoietic stem cells produce all types of blood cells by differentiating into specific blood cells in response to chemical signals. **Erythrocytes** are the most abundant blood cells, with approximately 5 million/mm³. These cells lack a nucleus and mitochondria, and each individual cell only survives approximately 120 days. Erythropoietin (EPO), a hormone produced by the kidney, stimulates erythrocyte production. Erythrocytes contain hemoglobin (Hb), a protein that contains an iron molecule (Fe) and transports both O_2 and some CO_2. The Fe in Hb binds O_2, while CO_2 binds other regions of the protein. Pathologies of erythrocytes or Hb can occur that decrease the ability to carry O_2. **Methemoglobin** is when the iron molecule in Hb is in the wrong form (ferric, Fe^{3+}) instead of the correct form (ferrous, Fe^{2+}). If the iron is in the ferric form, Hb cannot bind O_2. **Carboxyhemoglobin** occurs when

CO (carbon monoxide) binds to hemoglobin. CO is produced by a furnace or a car and binds to the same site (iron molecule) in Hb. Since CO has a higher affinity for Hb than O_2, it therefore displaces O_2 from Hb, even if CO is in lower concentrations than O_2. In this case, death can occur due to lack of O_2 delivery to the tissues. Anemia occurs when there is either 1) too few erythrocytes (low hematocrit) or 2) too little Hb present in each erythrocyte. In either case, the oxygen carrying capacity of the blood is reduced.

Platelets are also produced in the bone marrow. These are cell fragments from a megakaryocyte (large cell). These cell fragments are involved in blood hemostasis (~clotting) and are very small on a blood smear.

Leukocytes

Leukocytes are white blood cells that make up approximately 4,000-10,000 cells/mm^3 of blood. These cells are classified as granular or agranular. **Granular** cells contain lobed or segmented nuclei with granules in the cytoplasm. There are three types of granular cells: Eosinophils, Basophils, and Neutrophils.

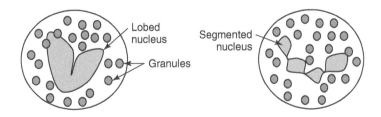

Eosinophils contain bright red granules in the cytoplasm and are approximately 10-14 μm in size. These cells are phagocytic, especially for parasites. They are increased in numbers in the blood in parasitic disease (intestinal worms, lice, scabies mites, etc.) or chronic allergic disease (hay fever, asthma, etc.). Eosinophils will migrate to tissues undergoing allergic reactions and are fairly rare (1-4% of leukocytes; 100-400/mm^3).

Basophils contain dark blue granules and are approximately 10-12 μm in size. These cells are related to tissue mast cells and are often involved in allergic reactions including asthma and anaphylactic shock (a severe and life-threatening reaction to an allergen). Basophils release histamine and contribute to tissue inflammation. They are very rare (<1% of leukocytes; 20-50/mm^3).

Neutrophils contain mostly a clear cytoplasm on routine stains. The granules are present but don't stain well. These cells are approximately 10-14 μm and have a life span of 2-3 days. These cells are phagocytic and are found in increased numbers in the blood during bacterial infections (called neutrophilia). Neutrophils can also be found in pus and are quite common (50-80% of leukocytes; 3,000-7,000/mm^3).

Agranular leukocytes contain round or kidney-shaped nuclei with no cytoplasmic granules. Monocytes and lymphocytes are two types of agranular leukocytes. **Monocytes** contain kidney-shaped nuclei with a ground glass appearance in the cytoplasm. These are large cells (14-24 μm) and are the most phagocytic cell. They leave the blood to differentiate into tissue macrophages and can also be found in pus. Monocytes are the "clean up" cell and can engulf foreign material and pathogens. These cells are moderately rare (2-8% of leukocytes; 100-700/mm^3).

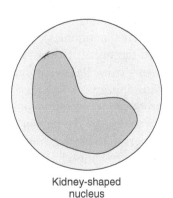

Kidney-shaped
nucleus

Lymphocytes contain a round nucleus with very little cytoplasm. These cells are small (5-17 μm) and are common (20-40% of leukocytes; 1,500-3,000/mm^3). **B-lymphocytes** (or B-cells) are produced and mature in the bone marrow. These cells are responsible for humoral immunity and produce antibodies that circulate in the blood. **T-lymphocytes** (or T-cells) are produced in the bone marrow, mature in the thymus (child) and bone marrow (adult). T-cells are responsible for cell-mediated immunity. These cells can kill virus- or bacteria-infected cells, cancer cells, and foreign tissue (transplant rejection).

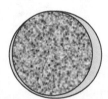

Round blotchy nucleus
with very little cytoplasm

Immunity

We have two types of immunity- nonspecific and specific immunity. **Nonspecific immunity** is mediated by all white blood cells <u>except</u> lymphocytes. Cells respond to any invader but the response is not very specific. The immune response is very fast, but short lasting. B- and T-lymphocytes are involved in **specific immunity**. There are millions of B- and T-cells in the body and each has a specific receptor that binds a specific invader antigen. Once the B- or T-cell binds an antigen, it becomes stimulated and divides repeatedly to form clones of itself (clonal selection). These clones can differentiate into two different cell types: effector cells and memory cells.

Effector cells are specific to the invader and are short lived (months to a few years). B-cell effector cells are **plasma cells**, which secrete antibodies specific to the invader. Antibody secretion begins within 10 days of the original infection and peak antibody production occurs around 21 days post infection. There are two types of T-cell effector cells: cytotoxic T-cells and helper T-cells. **Cytotoxic T-cells** (CD8 positive) are responsible for lysing infected body cells, and **helper T-cells** (CD4 positive) "help" stimulate the immune response. Helper T-cells are deficient in AIDS patients.

Memory cells are long lived (years to a lifetime) and form our lasting immunity. Both B- and T-cells can produce memory cells. **Vaccinations** are used to stimulate memory B- and T-cell formation in order to give an individual lasting immunity prior to pathogen exposure. After vaccination, individuals have large numbers of memory cells, which can then have a massive and very quick (within 2-7 days) response to an invading pathogen. When exposed a second time, memory B cells differentiate into plasma B- cells which produce/secrete large numbers of antibodies. These antibodies can then surround the invader and prevent its entry into cells. Often, an individual will not even display symptoms (upon exposure to invader) due to this rapid antibody production.

Specific immunity can be either active or passive. **Active immunity** occurs when <u>antigens</u> stimulate lymphocytes through either natural exposure or vaccination. This will activate B- and T-cells to form effector cells (plasma cells, cytotoxic T-cells, helper T-cells) and memory cells. This will provide long-lasting immunity. **Passive immunity** occurs when <u>antibodies</u> themselves are directly transferred to another person's plasma. For example, this occurs naturally from mother to infant prenatally through the placenta and somewhat through the milk (the infant is given maternal antibodies). Because the antigen is not present, no memory or plasma cells are stimulated or formed in the infant. However, maternal antibodies bind to the invader's antigens and prevent the invader from entering cells. Passive immunity lasts only a few months – essentially passive immunity is lost once the antibody disappears.

Autoimmunity is when an antibody produced by an individual inappropriately binds to the person's own tissue antigens, causing the immune system to destroy the person's own tissues. For example, <u>scarlet fever</u> is due to strep bacteria, but can also cause the person to produce antibodies that bind to heart antigens and cause heart damage. Similarly, <u>type 1 diabetes</u> is due to antibodies produced by the patient that destroy their own insulin-producing cells (beta cells) in the pancreas. <u>Rheumatoid arthritis</u> and <u>multiple sclerosis</u> are also common autoimmune disorders.

White Blood Cell Counts

White blood cell counts determine the percent of each cell type in the blood and can be used for diagnostic purposes. The <u>differential count</u> of each cell type is an estimation of the percentage of each cell type present in a blood sample. The <u>actual count</u> can be calculated by taking the differential count and multiplying by the total number of white blood cells. The actual count is more important than the differential count. For example, 50% of 2000 (1000) is a low number of neutrophils while 50% of 20,000 (10,000) is a high number. In today's lab, you will perform a differential count, and calculate the actual count for each type of leukocyte.

Experiment: Performing a Differential White Blood Cell Count

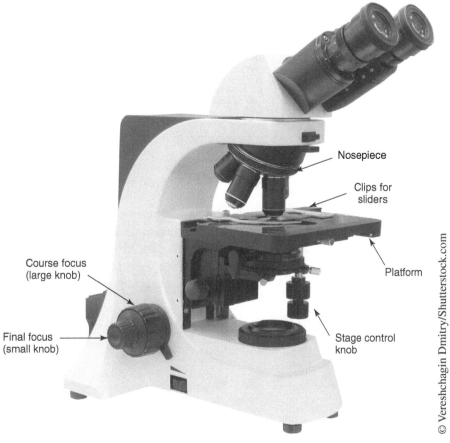

Figure 12.1

Procedure

1. Plug in the microscope at your station and turn the scope on (on/off switch on the right side of the scope).
2. Use the coarse focus adjustment knob to move the stage all the way down.
3. Place the slide on the stage (*i.e.*, the platform) and secure with the stage clip(s).
4. Turn the nosepiece to move the lowest power objective lens (4X) in place.
5. Bring the specimen on the slide into focus by using the using the course and/or fine focus adjustment knobs.

6. Repeat step 5 with the 10X objective lens.
7. Turn the nosepiece to move the 40X objective into place. Use only the fine focus adjustment knob to bring the specimen into focus when using this objective lens.
8. Once in focus with the 40X objective, turn the nosepiece to the right so that the 40X and the 100X objectives are on either side of the slide.
9. Place 1–2 drops of oil on the slide.
10. Rotate the nosepiece back to the left to bring the **100X** objective into the oil on the slide.
11. Focus using only the fine adjustment knob until the blood cells come into focus.
    Note:
    - If you accidently rotate the 40X objective into the oil, contact your instructor immediately to clean it off.
    - If you cannot get the specimen into focus suing the 100X objective lens, contact your instructor.
12. Use the stage control knob to move across the slide in a systematic manner, and count/record each cell type you see. Record these values on the data sheet at the end of this exercise.

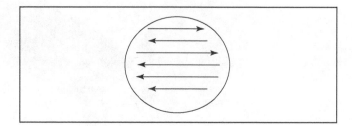

13. Identify and record a total of 50 white blood cells within your group (*i.e.*, ~ 17 white blood cells per person).
14. Determine the percentage of each cell type by dividing the counted number of each cell type by 50 (total number of counted cells) and then multiplying by 100. For example, if 35 neutrophils were counted within your group, you would perform the following calculation: $35 \div 50 = 0.7 \times 100 = 70\%$.
15. Determine the actual number of each cell type by multiplying the percentage of the total by the total white blood cell count. In the example provided in step 14, the actual number of present would be: $0.7 \times 10,000 = 7,000$.
16. Perform these calculations in the data sheet at the end of this exercise.

Lung Volumes & Capacities

Study Questions

1. Describe inhalation and exhalation in terms of Boyle's Law.
2. How is forced exhalation different from relaxed exhalation?
3. Why is intrapleural pressure always negative?
   a. What condition causes intrapleural pressure to become equal to atmospheric pressure?
   b. What happens to the lung?
4. What is surfactant? Why is it important?
   a. Under what conditions would you find a lack of surfactant?
5. Define:
   a. Tidal volume
   b. Inspiratory reserve volume
   c. Expiratory reserve volume
   d. Vital capacity (VC)
   e. Total lung capacity.
6. Why must we have a residual volume in the lungs? Under what condition does this volume leave the lung?
7. What is forced expiratory volume 1.0 ($FEV_{1.0}$)? Why is it important clinically?
8. What is obstructive disease?
   a. What lung volumes and capacities are affected?
   b. What conditions are considered obstructive diseases?
9. What is restrictive disease?
   a. What lung volumes and capacities are affected?
   b. What conditions are considered to be restrictive disease?
10. What is emphysema? What causes it? Is it a restrictive or an obstructive disease?

Applied Questions (Answers in back)

1. Josie thinks she may be having an asthma attack. What test would you use to determine if she has asthma? What change would you see in the test?
2. Bill has fallen on a sharp piece of metal that has cut an opening in his chest wall. What would happen to his intrapleural pressure? Is there any way you could help him breathe?

Introduction

The process of breathing involves the exchange of air between the environment and lung alveoli. To breathe, air has to move in and out of the lungs. Air moves as a result of pressure differences between the atmosphere and alveoli. Air always moves from higher pressure to lower pressure. **Boyle's Law of Gases** (at constant temperature) states that air pressure changes are inversely related to volume changes. For example, ↑ volume = ↓ pressure and ↓ volume = ↑ pressure. Essentially, if a gas is compressed (decreased volume), then pressure of the gas increases. Airflow is dependent on pressure (P) and resistance (R), such that flow = $\frac{\Delta P}{R}$ (just like the blood flow formula). High resistance to airflow occurs during disease states such as asthma, chronic bronchitis, and emphysema, which cause a decrease in airflow through the lungs.

The **intrapleural space** is a tiny space between the lungs and the wall of the thoracic cavity. Because the chest is a closed cavity (not open to air), and the lungs are elastic and tend to collapse, the pressure in the intrapleural space is always <u>negative</u> with respect to atmospheric pressure. The negative pressure (suction) opposes the elastic recoil of the lungs and helps keep the lungs from completely collapsing. To expand the lungs during **inspiration** (breathing in), the pressure in this space becomes even more negative. To allow the lungs to collapse a bit during **expiration** (breathing out), the pressure in this space becomes less negative. If the chest cavity is punctured, air is "sucked" into the thorax, causing the lungs to collapse. This is called a **pneumothorax.** If both sides of the chest are open to the atmosphere, both lungs will collapse and the patient cannot expand them with air.

<u>Inspiration</u> occurs when the diaphragm and external intercostal muscles contract, causing the thoracic space to expand. The increased volume of the thoracic space results in a decrease in air pressure in the intrapleural space, which causes the lungs to expand. As the lungs expand, the volume increases and pressure decreases within the alveoli. When alveolar pressure is less than atmospheric pressure, air moves into the lungs.

<u>Expiration</u> occurs when the diaphragm and external intercostal muscles relax (and internal intercostals contract), causing a decrease in the volume of the thoracic space. As the volume decreases, air pressure increases in the intrapleural space, causing the contraction of the lungs due to elastic recoil. The alveolar volume decreases and pressure increases, causing air to move out of the lungs (from higher pressure in the lungs to lower pressure in the external environment). Expiration is generally passive at rest; it requires little or no energy. During exercise or disease, heavy expiration will utilize the internal intercostals to affect chest cavity volume.

Summary of inspiration and expiration
1. Chest cavity always changes first.
2. When the thoracic volume changes, pressure changes in the intrapleural space.
3. Changes in lung volume cause pressure changes in the alveoli.
4. Pressure in the alveoli becomes either less than or greater than atmospheric.
5. Air moves from higher to lower pressure.

Surfactant is a liquid secreted by alveolar cells that reduces the surface tension of alveoli. Surface tension is the attraction between the walls of an alveolus that tend to make the alveoli collapse. Thus, surfactant helps keep the alveoli from collapsing. Some premature infants have respiratory distress syndrome, where the lungs are too underdeveloped to produce surfactant. The alveoli collapse and the infant cannot inflate the lungs.

Lung Volumes and Capacities

When measuring the volume of an individual's lung as they breathe normally and then take a deep breath, you would get a graph that looks something like this:

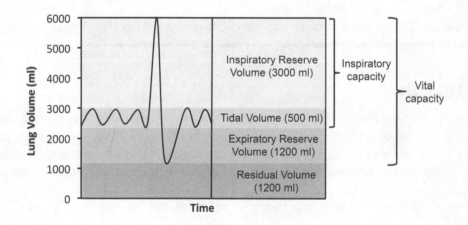

We can identify several key respiratory volumes from the graph above. The normal volume of air inspired or expired during one respiratory cycle under resting conditions is called the **tidal volume**. Tidal volume is normally ~600 ml (adult male) or ~500 ml (adult female). The volume of air that can be inspired *after* (in addition to) the completion of normal resting inspiration is the **inspiratory reserve capacity** (~ 2500–3000 ml). The **expiratory reserve capacity** is the volume of air that can be expired *after* (in addition to) the completion of a normal resting expiration (~1200 ml). The **total lung capacity** is the amount of air in the lungs following a forced maximal inspiration. The **residual volume** is the volume of air that remains in the lungs after a forceful expiration (~1200 ml). This volume keeps the lungs from collapsing unless the chest is punctured. The **vital capacity (VC)** is the maximum amount of air that can be forcefully expired following a maximal forced inspiration (*i.e.*, the maximum amount of air that can be moved through the lungs in a given breath). Values for VC depend on body size and physical conditioning. VC is an important diagnostic tool for pulmonary function, especially for restrictive lung disease. Another important diagnostic tool for pulmonary function is the **forced expiratory volume test 1.0 (FEV$_{1.0}$).** The FEV$_{1.0}$ is the percent of the VC that can be forcefully exhaled in 1 second. This is reduced in patients with obstructive lung disease and can be calculated by the following formula:

$$\text{FEV}_{1.0} = \frac{\text{Volume expired in 1 second}}{\text{VC}} \times 100\%$$

Lung Disease and Volumes

Obstructive lung disease is a disease that causes a high airway resistance and, thus, a slower airflow. This type of disease reduces FEV$_{1.0}$. Examples of obstructive disorders include: asthma, chronic bronchitis, and emphysema.

Restrictive lung disease causes less air to be moved into or out of the lungs. These diseases reduce VC. Examples of restrictive disorders include pulmonary fibrosis, where thickening and stiffening of the alveolar walls results in lungs that don't expand (stretch) properly to fill with air. Emphysema is also a restrictive disorder because the alveoli are destroyed and coalesce into large air pockets that do not exchange with atmospheric air. Air remains trapped in the lung and the amount of air that can be moved through the lungs is decreased.

In today's lab, you will measure your tidal volume, VC and FEV$_{1.0}$.

Experiment: Measuring Lung Volumes & Capacities

Setup

1. Log in to the computer at your station.
2. Turn the MP36 on using the switch at the back of the unit.
3. Double click on the BIOPAC BSL Student Lab 4.1 icon on the desktop.
4. In the **BSL Lessons tab**, select L-12 Pulmonary Function I and then click OK.
5. In the dialog box that appears, type in your section number followed by a space and then the year with no dashes or spaces (*i.e.*, Section 01 2017).
6. Click OK.
7. Enter your gender, age and height (in feet + inches) in the subject details dialog box that appears (**Figure 13.1**) and then click OK.

Calibration

1. In the next screen (see **Figure 13.2**), the Hardware tab below the empty graph window will be selected, and underneath the tab it should say 'Verify MP connections.'
2. Click on the Next Tab>> button at the bottom of the graph window (**Figure 13.2**) to take you to the Calibration tab.

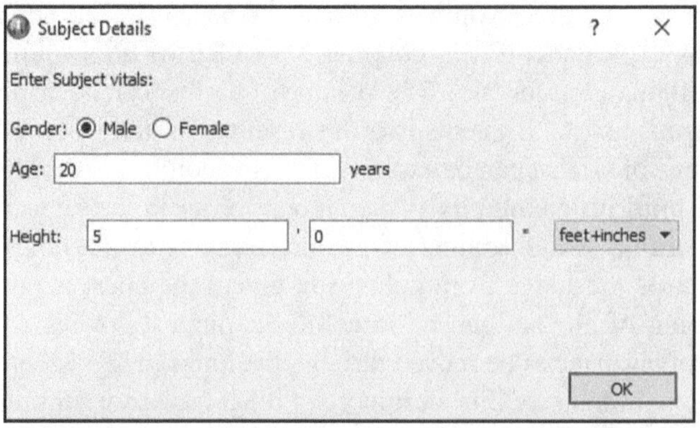

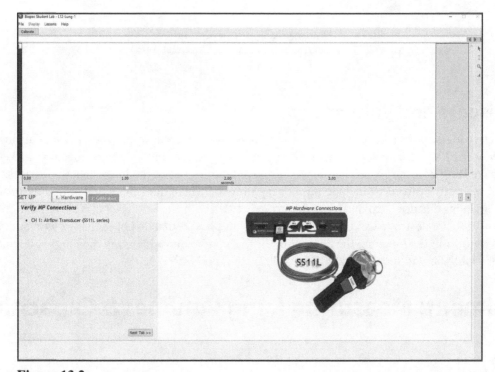

Figure 13.1

Figure 13.2

3. Click on the Calibrate button and hold the transducer still with the inlet side pointing toward the screen. The recording will run a line for two screens and stop by itself. Your data should resemble **Figure 13.3**.
4. If your data resembles that in Figure 13.3, click on the Continue button. If it doesn't, click on Redo calibration until it does and then press Continue.
5. The Hardware tab with Verify MP connections will show below the graph window. Click on the Next Tab>> button to take you to the Calibration tab.
6. Attach the calibration syringe to the inlet side of the airflow transducer via the blue coupler on the syringe (See **Figure 13.4**).
7. Once attached, pull the syringe plunger all the way out.
8. To calibrate:
   a. Hold onto the syringe with both hands; one hand on the plunger and the other under the blue coupler as shown in **Figure 13.4**. **DO NOT HOLD ON TO THE AIRFLOW TRANSDUCER!!**
   b. Push the syringe quickly in (~1 second of time).
   c. Wait 2 seconds, and pull the syringe quickly out (~ 1 second).

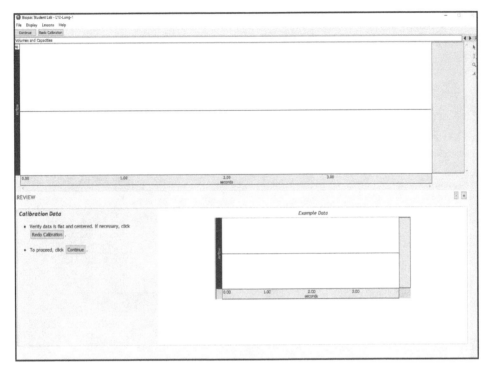

Figure 13.3

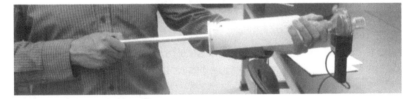

Figure 13.4

d. Wait 2 seconds and repeat four additional times for a total of 5 complete cycles or 10 total movements (See **Figure 13.5**)
e. Click End Calibration.
9. If your data resembles that in **Figure 13.5**, click Continue. If it doesn't, click on Redo. calibration until it does and then press Continue.
10. Remove the calibration syringe from the air filter.

A: Measuring Lung Volumes and FEV$_{1.0}$

1. Attach a sterile filter to the inlet side of the air flow transducer.
2. Click on the Next Tab>> at the bottom of the graph window.
3. Ignore the directions underneath the subject tab and follow those below.
4. The subject should hold his/her nose closed (or use a nose clamp) and breathe through the apparatus for about 30 seconds to get used to it.

NOTE:
• Ensure that no air leaks out around the air filter otherwise your measured volumes will be smaller than your real volumes.
• Be sure to keep the air flow transducer upright.
• The lung volumes will not appear while you are recording your breathing. The computer recording will show only air flow. Volumes will appear after you press stop.

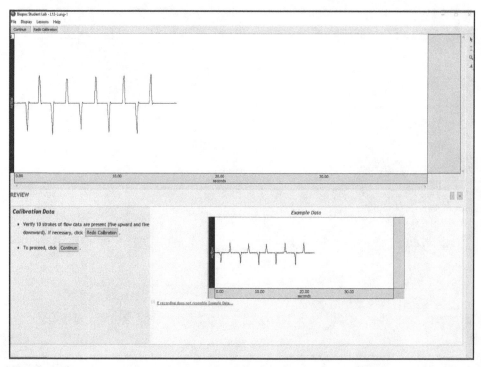

Figure 13.5

5. The controller should click Record button in the graph window.
6. The subject should:
   a. Breathe normally for three breaths.
   b. Inhale as deeply as he/she can and then hold his/her breath for 1 second.
   c. Blow the air out as fast as he/she can, blowing out as much air as possible.
   d. Breathe normally for three breaths.
7. Press Stop. Your data on channel 2 should resemble that in **Figure 13.6**.

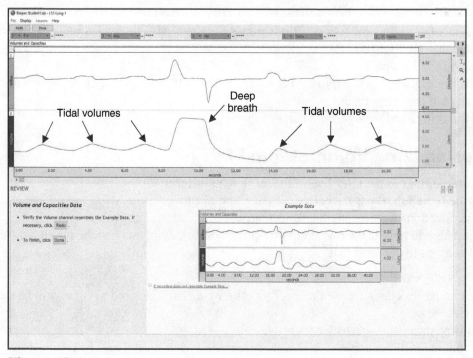

Figure 13.6

8. If you make a mistake, click Redo, and repeat steps 6 and 7.
9. If the data looks like that in **Figure 13.6**, click on Done.
10. The dialog box shown in **Figure 13.7** should appear. Select yes.

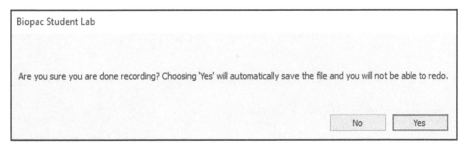

Figure 13.7

11. In the dialog box that appears (**Figure 13.8**), click on Analyze current data file and then OK.

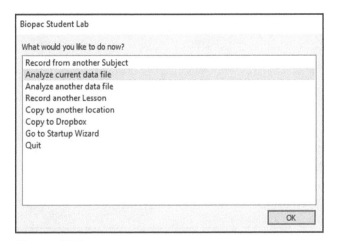

Figure 13.8

B: Data Analysis

1. Drag to highlight all of the written material in the journal and press Delete.
2. Adjust the second set of measuring settings at the top of the graph window. To do so, click on the max and select none from the drop down list.
3. Repeat with the next sets of boxes - changing the min to none, and then the delta to none.
4. Your settings should reflect those in presented in **Figure 13.9**

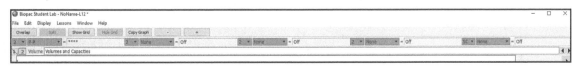

Figure 13.9

5. Click on the Zoom icon (magnifying glass) in the right screen and drag a box over two tidal volume waves, your VC/FEV$_{1.0}$, and a following tidal volume wave as shown in **Figure 13.10**. If your zoom was incorrect, choose Display >Zoom Previous.

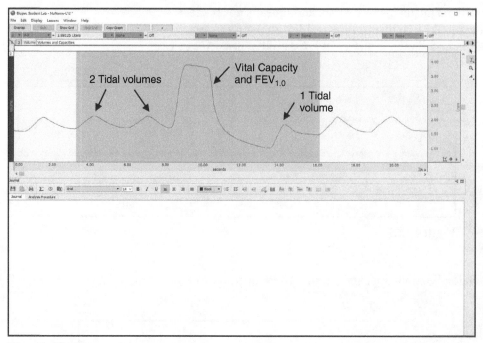

Figure 13.10

6. Click on the I–icon on the right side of the graph window and measure the following lung volumes/capacities:
   a. Tidal volume:
   - Click in the journal and type Tidal Volume and hit Enter.
   - Click on the I-icon to drag and highlight a section of the graph that includes the peak of a tidal volume wave and the lowest point (See **Figure 13.11**).

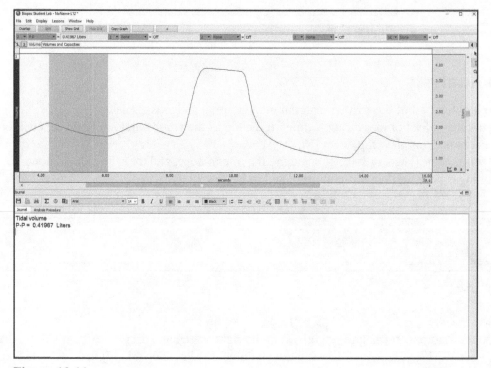

Figure 13.11

- Press Ctrl-M on your keyboard to record this data in your journal. The p-p value, which measures the difference between the highest and lowest points in the highlighted area, is your tidal volume.
- Record this value on your data sheet at the end of this exercise.

b. Vital Capacity:
- Click in the journal below your tidal volume data and type Vital Capacity and hit Enter.
- Using the I-icon, drag and highlight from the peak of your first deep breath to the lowest point of your maximum expiration (**Figure 13.12**).

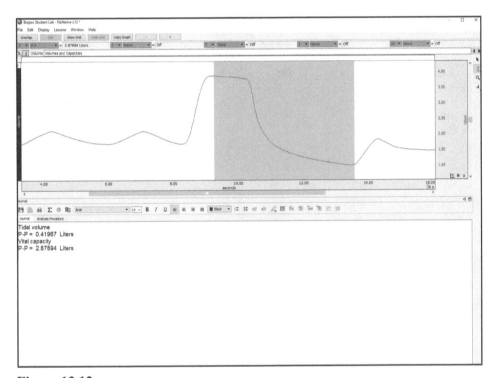

Figure 13.12

- Press Ctrl-M on your keyboard to record this data in your journal.
- Record this value on your data sheet at the end of this section.

c. Measuring $FEV_{1.0}$:
- Adjust the second set of measuring settings at the top of the graph window. To do so, click on the none and select delta T from the drop down list. Delta T will determine the time difference between the beginning and the end of the selected area.
- Click in the journal below your VC data and type $FEV_{1.0}$ and hit Enter.
- Type expired in one second and hit Enter.
- Locate the VC/$FEV_{1.0}$ wave.
- Using the I-icon, drag and highlight from the beginning of the rapid expiration portion of the wave (the point where the line begins to drop) until the delta T box at the top of the screen reads as close to 1.000 seconds as you can get (**Figure 13.13**).
- Press Ctrl-M on your keyboard to record this data in your journal.
- Record this value on your data sheet at the end of this section. Use it to calculate your $FEV_{1.0}$ as described on the data page at the end of this section.

7. Save your data (File>Save).
8. Turn in the data sheet at the end of this exercise to your instructor at the end of your lab session.
9. Close the MP36 graph window and turn the Biopac off via the switch in back of the MP36 unit.

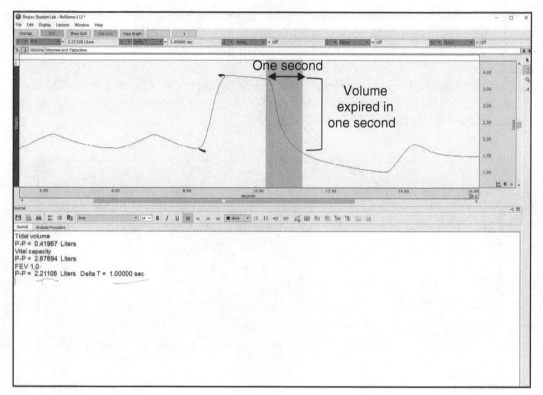

Figure 13.13

Regulation of Respiration

Study Questions

1. In the negative feedback control of respiration, what is the sensor? The integrator? The effector?
2. How is blood pCO_2 related to H^+?
3. How does hypoventilation cause acidosis?
4. How can respiration cause an alkalosis?
5. During hyperventilation, predict the values of each of the following when compared to normal respiration:
   a. alveolar pCO_2
   b. alveolar pO_2.
6. Just after a period of hyperventilation, predict the respiratory rate and depth when compared to normal.
   a. Why did this change in respiration occur?
   b. Did the change in CO_2 or O_2 have the greatest effect? Why?
7. Why does a small decrease in blood pO2 have no effect on respiratory rate or depth? What effect does a large decrease in pO_2 have? Which chemoreceptor is affected?
8. When rebreathing into a bag, what happens to the following when compared to breathing normally?
   a. Alveolar pCO_2
   b. Alveolar pO_2
   c. Predict the respiratory rate and depth when compared to normal. Explain your answer.
9. Predict the effect of increased anatomical dead space on respiratory rate and depth when compared to normal. Explain your answer.
10. Explain the effect of exertion on breathing. Is this due to a change in blood pO_2 or pCO_2?

Applied Questions (Answers in back)

1. Calamity Jane has fallen off her bike and cracked several ribs. This has made respiration painful and she is not breathing well.
   a. What is this condition called?
   b. Predict the following values when compared to a normal individual:
      1) alveolar pCO_2
      2) blood pCO_2
      3) alveolar pO_2
      4) blood pO_2.
   c. What would you predict about her blood pH? Explain.
   d. If she was given a higher oxygen content in her inhaled air, would this correct her blood pO_2? Her blood pCO_2? Her blood pH?
2. Carbon monoxide poisoning is treated with inhaled air that is high in oxygen and also has CO_2 added. Why might this be beneficial in eliminating carbon monoxide from the body?

3. Randy Renal has a kidney infection. He is unable to eliminate all the acid from his body, and he has a metabolic acidosis–a blood pH that is low (high level of H^+ in the blood).
   a. What would you predict about his respiratory rate and depth? Explain.
   b. What would you predict about his blood pCO_2? Explain.

Introduction

Breathing, including respiratory rate and depth, is regulated by a negative feedback system. Arterial blood pCO_2 and pO_2 are strongly correlated with those concentrations in the lung alveoli. The body breathes in O_2 during each inspiration and eliminates CO_2 during every expiration. The sensor in this system is the chemoreceptors located in the arterial system and the medulla. Chemoreceptors detect CO_2, H^+ and O_2 levels in the arterial blood and send information to the **respiratory control center** in the medulla (integrator), which makes appropriate adjustments to the respiratory rate and depth (minute ventilation) by modifying contractions of the diaphragm and respiratory muscles (effectors). **Minute ventilation** is the amount of air moved in or out of the lungs in one minute, and is determined by multiplying breaths per minute (rate) by volume of each breath (depth).

CO_2 and H^+ levels are more important in regulating our minute ventilation than O_2 levels. An enzyme called **carbonic anhydrase** catalyzes conversion of CO_2 and water into bicarbonate and H^+ such that:

$$CO_2 + H_2O \leftrightarrow H_2CO_3 \leftrightarrow HCO_3^- + H^+$$

If too much CO_2 is present, more H^+ ions are formed and the blood pH becomes lower (shifts equation to the right):

$$CO_2 + H_2O \rightarrow H_2CO_3 \rightarrow HCO_3^- + H^+$$

If not enough CO_2 is present, H^+ ions combine with bicarbonate to form more CO_2, causing blood pH to rise (shifts equation to the left):

$$CO_2 + H_2O \leftarrow H_2CO_3 \leftarrow HCO_3^- + H^+$$

Thus, the regulation of blood pH is closely tied with CO_2 levels and respiration. **Peripheral chemoreceptors**, located in the carotid and aortic bodies, detect the carbon dioxide (pCO_2), hydrogen ion (H^+) and oxygen (pO_2) levels in arterial blood. **Central chemoreceptors** are located in the medulla oblongata and these sense only hydrogen ion (H^+) concentrations (through the conversion of CO_2 to bicarbonate and H^+) in arterial blood. Chemoreceptors are more sensitive to changes in pCO_2 and H^+ levels than they are to pO_2. A small change in pCO_2 and (H+) immediately results in a change in respiration. However, a large drop in pO_2 level is needed to alter respiration. Increased PO_2 has very little effect on respiration. Increased pCO_2 and H^+ will stimulate respiration (increase respiratory rate and depth), while decreased pCO_2 and H^+ will inhibit respiration (decrease respiratory rate and depth).

Today's lab will examine different situations that alter breathing rate and depth. For each exercise, you will predict the effect of the manipulation on CO_2, H^+ and O_2 levels in the arterial blood. Then you will predict the respiratory control center's corrective response (i.e. increased or decreased minute ventilation). We will examine 4 different situations: hyperventilation, rebreathing in a bag, breathing in a tube, and exertion.

Hyperventilation

Hyperventilation (breathing too rapidly and deeply) will eliminate too much CO_2 from the lung alveoli, and thus will reduce pCO_2 in the blood. Lowering blood pCO_2 reduces the blood H^+ levels, increasing blood pH. This causes the chemoreceptors to signal the respiratory control center, which will reduce respiratory rate and depth. Decreased minute ventilation will bring CO_2 (and H^+) levels back to normal. Hyperventilation does not have much effect on blood pO_2.

Rebreathing in a Bag

When rebreathing into a bag, an individual is breathing in <u>used/stale</u> air (high CO_2 and low O_2). This increases the blood pCO_2 and H^+ concentrations, and decreases blood pO_2. Chemoreceptors will signal the respiratory control center that pCO2 and H^+ is elevated, and PO_2 is decreased. The respiratory control center will respond by increasing minute ventilation to get rid of excess CO_2 and to increase O_2.

Increasing Anatomical Dead Space: Breathing through a Tube

Anatomical dead space is the space inside the conducting airways (nasal passages, trachea, bronchi and bronchioles). The air in this space does not exchange gases with the blood. At the end of each inspiration, air in the airways is relatively fresh air (similar to outside or atmospheric air – high pO_2, low pCO_2) - although it has mixed somewhat with the anatomical dead space air. Blood gases exchange in the alveoli and, by the end of each expiration, air in the conducting airways is similar to alveolar air (higher pCO_2, lower pO_2). This "leftover" air in the airways (used/stale) is <u>rebreathed</u> into the alveoli during every normal inspiration and our body has evolved to utilize this somewhat "mixed" air. However, breathing through a long tube (Ex. snorkel) artificially lengthens the dead space, and therefore increases the volume of used/stale air that is rebreathed with each breath. This leads to an increased CO_2, which subsequently signals for an increased minute ventilation in an attempt to compensate.

Exertion

During exertion (exercise), cells such as muscle and cardiac cells produce more CO_2 and require more O_2. Yet, arterial blood levels of pCO_2 and pO_2 **do not change** during exercise. Venous blood gases **do** change because they reflect tissue activity. Information from the mechanoreceptors and higher cortex (brain) is sent to the respiratory center and result in an increase in respiratory rate and depth. This increase in minute ventilation allows the body to meet its additional O_2 needs and eliminate the extra CO_2 produced in the exercising cells. The respiration increase during exercise occurs due to a **feed forward mechanism.** The increase occurs in anticipation of a need – preventing changes in arterial pO_2 and pCO_2. Following exertion, the respiratory rate and depth remain elevated for a period of time known as oxygen debt.

Setup

1. One member of your group should perform each of the following functions:
   a. **Director:** Reads the directions.
   b. **Controller:** Runs the computer.
   c. **Subject:** person from whom data are being recorded.
2. Log in to the computer at your station.
3. Turn the MP36 on using the switch at the back of the unit.
4. Double click on the BIOPAC BSL Student Lab 4.1 icon on the desktop.
5. Click on the PRO Lessons tab, select Respiration Control and then click OK.

Subject Setup

1. Using the bulb, pump air into the respiratory cuff to partially fill it.
2. Place the respiration belt (with attached respiratory cuff) snugly around the upper part of the subject's abdomen, below the sternum and over the lower ribs.
3. The subject must sit in a chair with his/her arms resting on the table top, back straight, and feet resting on the circular foot support of the chair or the floor.

Note:
- The partially filled respiratory cuff should be positioned just under the lower edge of the sternum, with Velcro out and tubes exiting in a downward direction.
- The respiratory belt should be worn over as few layers of clothing as possible (T-shirt is okay) and should be snug, but not uncomfortable.
- The subject must sit still and **NOT** change position during the data collection, as changes in position affect the respiratory waves.

Calibration

1. Click on Start.
2. The subject should breathe normally for 3 breaths.
3. He/she should then take a deep breath and blow all the air out.
4. The subject should breathe again normally for 3 breaths.
5. Your results should resemble those in **Figure 14.1**. If they don't, contact your instructor.
6. Save your data (File>Save as) in your section's folder within the Student Data folder on the desktop. Name your file Respiration Control.

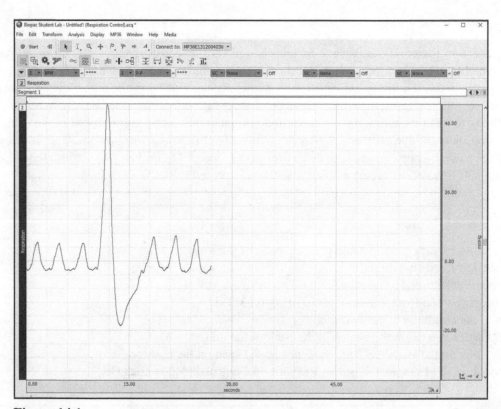

Figure 14.1

Experiment 1: Hyperventilation

Procedure

1. Insert a marker event at the end of the calibration data by right clicking in the bar below the event bar. In the dropdown menu that appears, click on Insert New Event.
2. Right click on the marker triangle and click Edit event. Highlight and delete any writing in the marker text box, and type 'hyperventilation-control' (See Exercise 3, Figure 3.2).
3. Click on Start. Record a normal resting breathing pattern for at least 6–8 breaths. **DO NOT STOP RECORDING!!**
4. Insert a marker (push the Esc key) just as the subject starts to hyperventilate. He/she should continue until a noticeable change in his/her breathing is observed on the screen or he/she feels slightly dizzy.
5. Insert a marker (click on the Esc key) just as the subject stops hyperventilating.
6. The subject should then relax and breath normally, letting his/her normal respiratory control take over. Record his/her breathing until the waves appear normal (~ 60 –90 seconds).
7. Press Stop.
8. Scroll back and label each of the inserted markers (**Figure 14.2**).
   - The marker after the 'hyperventilation-control' marker should be labeled 'begin hyperventilation'.
   - The next marker, the one inserted when the subject stopped hyperventilating, should be labeled 'end hyperventilation'.
9. Save your data (File>Save).

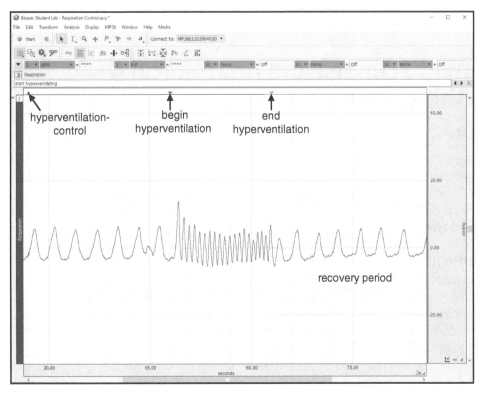

Figure 14.2

Experiment 2: Rebreathing in a Bag

Procedure

1. Insert a marker event at the end of the recovery period for experiment 1 by right clicking in the bar below the event bar. In the dropdown menu that appears, click on Insert New Event.
2. Right click on the marker triangle and click Edit event. Highlight and delete any writing in the marker text box, and type 'Rebreathing-control' (See Exercise 3, Figure 3.2).
3. Click on Start. Record a normal resting breathing pattern for at least 6–8 breaths. **DO NOT STOP RECORDING!!**
4. The subject should shake a plastic bag to fill it with air, place it over his/her nose and mouth and then begin breathing normally (trying not to control his/her breathing) into the bag.
5. Insert a marker (push the Esc key) just as the subject starts to breathe into the bag. He/she should continue until a noticeable change in his/her breathing is observed on the screen or if nausea or dizziness occurs.
6. Insert a marker (click on the Esc key) just as the subject stops breathing into the bag.
7. The subject should then relax and breath normally, letting his/her normal respiratory control take over. Record his/her breathing until the waves appear normal (~ 60 –90 seconds).
8. Press Stop.
9. Scroll back and label each of the inserted markers.
   - The marker after the 'rebreathing-control' marker should be labeled 'begin rebreathing'.
   - The next marker, the one inserted when the subject stopped breathing into the bag, should be labeled 'end rebreathing'.
10. Save your data (File>Save).

Experiment 3: Increasing Anatomical Dead Space by Breathing through a Tube

Procedure

1. For this experiment, the subject will fasten a breathing filter to one end of the respiratory tube.
2. Insert a marker event at the end of end of the recovery period for experiment 2 by right clicking in the bar below the event bar. In the dropdown menu that appears, click on Insert New Event.
3. Right click on the marker triangle and click Edit event. Highlight and delete any writing in the marker text box, and type 'dead space-control' (See Exercise 3, Figure 3.2).
4. Click on Start. Record a normal resting breathing pattern for at least 6–8 breaths. **DO NOT STOP RECORDING!!**
5. Insert a marker (push the Esc key) just as the subject starts to breathe normally (trying not to control his/her breathing) through the tube. His/her nose should be held closed with the thumb and index fingers of the non-dominant hand. He/she should continue to breathe through the tube until a noticeable change in his/her breathing is observed on the screen or if nausea or dizziness occurs.
6. Insert a marker (click on the Esc key) just as the subject stops breathing into the bag.
7. The subject should then relax and breath normally, letting his/her normal respiratory control take over. Record his/her breathing until the waves appear normal (~ 60 –90 seconds).
8. Press Stop.
9. Scroll back and label each of the inserted markers.
   - The marker after the 'dead-space-control' marker should be labeled 'begin tube'.
   - The next marker, the one inserted when the subject stopped breathing into the tube, should be labeled 'end tube'.
10. Save your data (File>Save).

Experiment 4: Exertion

Procedure

1. Insert a marker event at the end of end of the recovery period for experiment 3 by right clicking in the bar below the event bar. In the dropdown menu that appears, click on Insert New Event.
2. Right click on the marker triangle and click Edit event. Highlight and delete any writing in the marker text box, and type 'exertion-control' (See Exercise 3, Figure 3.2).
3. Click on Start. Record a normal resting breathing pattern for at least 6–8 breaths. **DO NOT STOP RECORDING!!**
4. Insert a marker (push the Esc key) just as the subject starts to run in place. He/she should continue to run he/she is breathing heavily.
5. Insert a marker (click on the Esc key) just as the subject stops running.
6. The subject should then relax and breath normally, letting his/her normal respiratory control take over. Record his/her breathing until the waves appear normal (~90 – 120 seconds).
7. Press Stop.
8. Scroll back and label each of the inserted markers.
   - The marker after the 'exertion-control' marker should be labeled 'pre-exertion'.
   - The next marker, the one inserted when the subject stopped running, should be labeled 'post-exertion.
9. Save your data (File>Save).

Data Analysis: Measuring Respiratory Rate and Wave Amplitude

A. Hyperventilation

1. Click on the second icon from the left under the Start menu and then on journal in the drop down menu that appears (See Exercise 1, Figure 1.5).
2. In the journal window that appears at the bottom of the graph window, type hyperventilation and hit Enter.
3. Find your 'hyperventilation-control' marker using the left and right arrows (◄◄►►) on the right hand of the screen.
4. Measure a typical single wave in the control section. To do so, scroll to the right to find a relatively stable section of waves after the 'hyperventilation-control' marker but before the 'begin hyperventilation' marker. Click on the I-icon to drag across an entire respiratory wave, *i.e.*, from the beginning of the wave to the beginning of the next wave (**Figure 14.3**). Be sure to include any space between the waves.
5. Press Ctrl-M on your keyboard. Two values, respiratory rate (first value) and wave amplitude, an indicator of respiration depth (second value), will appear in your journal. Record these values on the data sheet at the end of this exercise.
6. Repeat step 5 to measure a typical single wave early in the recovery portion (*i.e.*, right after your end hyperventilation marker) of the hyperventilation exercise. You do <u>NOT</u> need to measure a wave during the actual period of hyperventilation.
7. Record these values on the data sheet at the end of this exercise.

B. Rebreathing in a Bag

1. Type rebreathing (below the hyperventilation data) in the journal window at the bottom of the graph window and hit Enter.
2. Find your 'rebreathing-control' marker using the left and right arrows (◄◄►►) on the right hand of the screen.
3. Measure, as described in part A, a typical wave in the 'rebreathing-control' section.

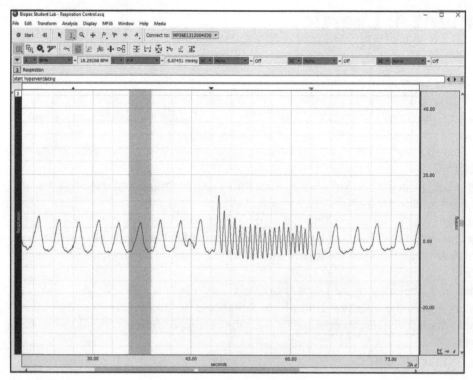

Figure 14.3

4. Repeat for typical waves near the ends of the 'begin rebreathing' (shortly before your 'end rebreathing' marker) and recovery period (after your 'end rebreathing' marker) sections of your data.
5. Record these values on the data sheet at the end of this exercise.
6. Save your data (File>Save).

C. Increasing Anatomical Dead Space by Breathing through a Tube

1. Type dead space (below the rebreathing data) in the journal window at the bottom of the graph window and hit Enter.
2. Find your 'dead space-control' marker using the left and right arrows (◄►) on the right hand of the screen.
3. Measure, as described in part A, a typical wave in the 'dead space-control' section.
4. Repeat for typical waves near the ends of the 'begin tube' (shortly before your 'end tube' marker) and recovery period (after your 'end tube' marker) sections of your data.
5. Record these values on the data sheet at the end of this exercise.
6. Save your data (File>Save).

D. Exertion

1. Type exertion (below the dead space data) in the journal window at the bottom of the graph window and hit Enter.
2. Find your 'exertion-control' marker using the left and right arrows (◄►) on the right hand of the screen.
3. Measure, as described in part A, a typical wave in the 'exertion-control' section.
4. Repeat for a typical wave early in the 'post-exertion' (shortly after your 'post-exertion' marker) section of your data.
5. Record these values on the data sheet at the end of this exercise.
6. Save your data (File>Save).
7. Close the MP36 graph window and turn the Biopac off via the switch in back of the MP36 unit.

Kidney/Urinalysis

Study Questions

1. What is the functional unit of the kidney?
2. What are each of the major divisions of the tubular system? What is the main function of each?
3. What substances are cotransported with sodium in the proximal tubule?
4. How is water reabsorbed? How is water reabsorption related to sodium reabsorption?
5. How is a dilute urine formed? What part of the tubular system is important for this? Why doesn't water follow sodium reabsorption in this segment?
6. Which part of the tubular system is responsible for adjusting sodium, water, potassium, and hydrogen reabsorption to meet the body's needs?
7. Which part of the tubular system can reabsorb free water (without reabsorbing Na^+ first)? What hormone is necessary for this to occur?
8. What is a transport maximum Tm? How does diabetes mellitus exceed the Tm for glucose? Where is glucose usually reabsorbed?
9. How does protein end up in the urine? Why is protein usually found in the urine only in very small amounts? What does a large amount of protein in the urine indicate?
10. What does glucose in the urine indicate? Why is glucose not usually found in the urine?
11. How is specific gravity (SG) measured? What does a high urine SG indicate? What about a low urine SG?
12. What is anti-diuretic hormone (ADH)? Where is it released? What stimulates its release?
13. What is diabetes insipidus? What hormone is deficient?
14. How does the urine from a patient with diabetes mellitus differ from the urine of a patient with diabetes insipidus?
15. What does the presence of ketones in the urine indicate? Is this always a serious condition?
16. What does hemoglobin in the urine indicate?
17. What disorders are associated with bilirubin in the urine?

Applied Questions (Answers in Back)

1. Joe has had the stomach flu for 2 days and is very dehydrated. Predict Joe's ADH level (high, low, or normal). Predict Joe's urine volume and SG (high, low, or normal).
2. Billie has a urine SG of 1.008, yet he is dehydrated. What hormone might Billie be unable to produce?

Introduction

Kidney

The nephron is the functional unit of the kidney. In the nephron, active transport of solutes creates a concentration gradient for water reabsorption by osmosis. Reabsorption of water concentrates urea and urea is reabsorbed by

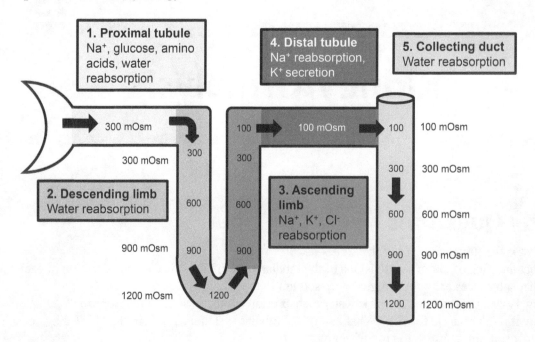

passive diffusion. Reabsorption is **unregulated** in the proximal tubule, descending limb and ascending limb of the loop of Henle. Reabsorption is **regulated** in the distal tubule and collecting duct to meet the needs of your body.

The **proximal tubule** contains the filtrate from the glomerulus. This is a site of <u>unregulated</u> reabsorption of many types of filtrate molecules (Ex. glucose, amino acids, ions, water) back into the bloodstream. Under normal conditions, the filtrate contains everything from the plasma except proteins and cells, as these are too large to leave the capillaries. Similar to plasma, the osmolarity of the filtrate in the proximal tubule is 300 mOsm and <u>does not change</u> as solutes are transported out because aquaporins allow water to move down its concentration gradient. For example, <u>all (100%)</u> of the glucose and amino acids in the filtrate are reabsorbed in the proximal tubule using Na^+ cotransporters.

The **descending limb of the loop of Henle** is only permeable to water (NOT solutes). Water follows its concentration gradient and is reabsorbed until the filtrate osmolarity is the same as the interstitial osmolarity. The osmolarity of the filtrate <u>increases</u> as fluid moves through the descending limb. This process is <u>unregulated</u>.

The **ascending limb of the loop of Henle** is only permeable to $Na^+/K^+/Cl^-$ (NOT water). Ions are actively pumped out of the filtrate, leaving water behind. The osmolarity of the filtrate <u>decreases</u> as fluid moves through the ascending limb. The filtrate osmolarity is 100 mOsm as it leaves the ascending limb, allowing for the formation of dilute urine. This process is <u>unregulated</u>.

The **distal tubule** is where <u>regulated</u> reabsorption of Na^+ and secretion of K^+ occurs to meet the body's needs. Aldosterone, a hormone produced by the adrenal gland, increases the amount of Na^+ reabsorption and K^+ secretion in the distal tubule.

The **collecting duct** is where <u>regulated</u> reabsorption of water occurs to meet the body's needs. This is under the control of anti-diuretic hormone (ADH), a hormone released by the pituitary gland of the brain. If ADH is present, the collecting duct is permeable to water (contains aquaporins) and water is reabsorbed by osmosis. Filtrate osmolarity increases as fluid moves through the collecting duct, resulting in small volumes of concentrated urine. If there is no ADH present, aquaporins are removed from the membrane and water cannot be reabsorbed. Osmolarity of the filtrate does not change, resulting in large volumes of dilute urine.

There are two forms of diabetes: diabetes insipidus and diabetes mellitus. In **diabetes insipidus**, patients cannot produce ADH. Therefore, the collecting duct is not permeable to water and they cannot produce concentrated urine. In **diabetes mellitus**, patients have a high glucose concentration in the plasma. More glucose than normal is filtered into the nephron and the glucose transporters in the proximal tubule are saturated. Some glucose, therefore, remains in the filtrate after the proximal tubule. Since there are no glucose transporters after the proximal tubule, glucose cannot be reabsorbed and it is eliminated in the urine. Glucose increases the osmolarity of the filtrate, altering water reabsorption throughout the nephron and resulting in large volumes of concentrated urine.

Urinalysis

Specific gravity (SG) is a measure of the concentration of a solution compared to pure water (SG = 1.000). For example, plasma has a specific gravity of 1.012, which indicates that plasma is more concentrated with solutes than pure water. Urine can range from a specific gravity of 1.010 (dilute) to 1.030 (concentrated). We measure specific gravity using a **urinometer**. This device floats in urine at different levels depending on the concentration of solutes in the urine. For example, more concentrated urine has a higher density of solutes, causing the device to float higher in the urine. We read the number indicated on the urinometer to three decimal places.

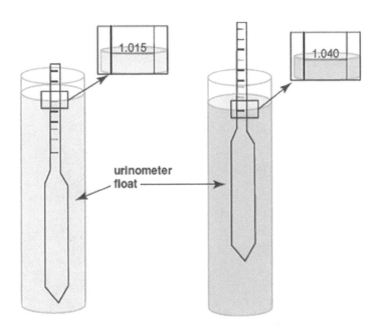

If we find that the urine sample has a high SG, then it is possible that the patient produced large amounts of ADH and was able to reabsorb most of the water in their filtrate. This could indicate dehydration. If we find that the urine sample has a low SG, then it is possible that the patient produced little or no ADH and water could not be reabsorbed (lost in the urine). This could indicate that the patient drank too much water, consumed alcohol (inhibits ADH release), or may have diabetes insipidus.

We can also measure compounds in the urine, such as glucose and protein, to diagnose patients. Both glucose and protein are not normally present in the urine of healthy individuals. If there is glucose present in the urine (glucosuria), this indicates that the patient has very high blood glucose levels and has diabetes mellitus. If there is protein in the urine (proteinuria), this indicates glomerular nephritis (glomerular damage). Because the glomerulus is damaged, the capillary is leaking protein into the nephron. Other compounds in the urine can also indicate disorders. For example, bilirubin in the urine (bilirubinuria) may indicate liver disease, hemoglobin in the urine (hemoglobinuria) indicates the excess destruction of erythrocytes, and ketones in the urine (ketoneuria) indicates excess fat metabolism.

In today's lab, you will analyze the urine from five patients and, using the results of your analysis, diagnose them with different disorders.

Experiment: Urinalysis

Procedure

1. Observe the color (pale yellow, yellow, or dark yellow) of the urine sample
2. Determine the volume of the sample in milliliters (mL)
   - To do so, remove the urinometer float from the sample, read the volume level at the meniscus and then replace the urinometer in the sample
   - Be sure to pay attention to the number of cylinders present at each station
3. Measure the specific gravity of the urine sample using with the urinometer float
   - Specific gravity = the number (between 1.000 and 1.060) on the urinometer where the top of the urine sample intersects with the float column
4. Perform a dipstick analysis
   - Aspirate a small amount of the unknown urine into the pipette at each station
   - Apply a small drop to each of the colored squares on the dip stick and allow it to sit for 60 seconds
   - Compare the square color for each test to the colors on the dipstick bottle
5. Compare the patient samples to the one of the normal urine samples set up on each side of the laboratory
6. Diagnose patients 1-5 as having one of the following disorders:
   - normal
   - dehydrated
   - diabetes insipidus
   - diabetes mellitus
   - glomerular damage (glomerulonephritis)

Urinalysis Data Sheet

Group Number: _____   Name: _____

Test	Normal	Patient 1	Patient 2	Patient 3	Patient 4	Patient 5
SG						
Color						
Volume						
Glucose						
Protein						
Diagnosis						

Possible Disorders: Normal, dehydrated, diabetes insipidus, diabetes mellitus, glomerular damage (glomerulonephritis)

Answers to Applied Questions

Homeostasis

1. Plasma level of thyroid hormone (TH) is controlled by negative feedback. The release of TH is stimulated by another hormone from the pituitary gland known as TSH. The loop is as follows:

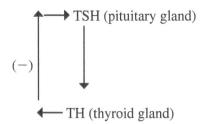

A patient with hypothyroidism (TH) due to a defective thyroid gland would likely have:
a. A very high TSH
b. A very low TSH
Explain your choice.

Answer: A very high TSH. Loss of the negative feedback of TH would result in a disinhibition of the secretion of TSH, and thus TSH levels would rise. This hormone is used clinically along with TH to diagnose thyroid deficiencies and also to regulate medication to the appropriate level of TH. TSH levels are supposed to be kept slightly high so that if the thyroid gland recovers, it will be stimulated to secrete TH, rather than be depressed by high supplementary (pharmaceutical) levels.

2. Jill Jones is sick with Strep. She has a very high temperature and is shivering.
   a. Why is she shivering instead of sweating?

Answer: Her setpoint for temperature regulation has been moved to a point that is higher than her present body temperature. Thus, according to her hypothalamus, she is still below her setpoint.

   b. Has her negative feedback system for temperature regulation failed?

Answer: No, her system is working properly.

3. Just as we begin to exercise, <u>before</u> our body's needs have increased, our respiration and heart rate get faster. What type of mechanism is this?

Answer: Feed forward.

Diffusion & Osmosis

1. Perry Pathologist just placed a slide of cells under the microscope. He discovered that all the cells appear swollen, and some have burst. He must have washed the cells with the wrong solution. Was the solution hypertonic, hypotonic, or isotonic? Explain.

Answer: Hypotonic.

1. Carrie's boss has told her to mix up an isotonic (290 mOsm) salt solution to rinse blood cells with. Carrie cannot remember how many grams of salt to add to 1 liter of water. Can you help her?

Answer:

$$= \frac{300 \text{ milliosmoles}}{\text{liter}} \times \frac{1 \text{ osmole NaCl}}{1000 \text{ milliosmoles}} = \frac{0.300 \text{ osmoles NaCl}}{\text{liter}}$$

$$= \frac{0.300 \text{ osmoles}}{\text{liter}} \times \frac{1 \text{ mole NaCl}}{2 \text{ osmoles NaCl}} = \frac{0.150 \text{ moles NaCl}}{\text{liter}}$$

$$= \frac{0.150 \text{ moles NaCl}}{\text{liter}} \times \frac{58.5 \text{ g NaCl}}{\text{mole NaCl}} = \frac{8.78 \text{ grams NaCl}}{\text{mole NaCl}}$$

Osmosis calculations: Complete the chart below.

grams/dl	Molarity	Osmolarity	Milliosmolarity	Effect on a cell (300 mOsm)	Tonicity (hyper-, iso-, hypo)
4 g NaCl/dl	0.684	1.368	1368	Shrink	Hyper
3 g NaCl/dl	0.513	1.026	1026	Shrink	Hyper
3 g glucose/dl	0.167	0.167	167	Swell	Hypo
0.5 g NaCl/dl	0.085	0.170	170	Swell	Hypo

2. A solution of NaCl has an milliosmolarity of 200. How many grams of NaCl were added to a 1 liter container of pure water?

Answer: 200 mOsm = 0.200 Osm = 0.1 molar NaCl = 5.85 g NaCl/liter.

Action Potential

1. The sensory nerves carry information from the skin to the brain. John touched his sister's arm to get her attention. She ignored him, so he punched her arm. From any <u>single</u> sensory nerve from the affected area, compare the frequency of action potentials from the touch to that of the punch.

Answer: Frequency of action potentials is greater with a stronger stimulus. Thus, the punch would stimulate a greater frequency of action potentials in each of the nerves than the touch.

2. Certain medications may make the resting membrane potential of a <u>single</u> nerve more negative. How might this affect the voltage that would need to be applied to bring the nerve to threshold (larger, smaller, or unchanged)? Would this affect the size of the action potential?

Answer: Since the nerve would need a greater change in membrane potential to get from the resting potential to threshold, a greater stimulus voltage must be applied. The size of the action potential should be unchanged once threshold is reached.

Special Senses

1. Tory Tense has muscle spasms in her left shoulder. She also has tingling in her left hand, yet her left hand appears normal. Is the problem in her hand, or in her shoulder? Explain.

Answer: Muscle spasms in the shoulder can be severe enough to put pressure on the sensory nerves along their route to the spinal cord, and create action potentials in these nerves. These arrive at the brain and are interpreted as signals (tingling) originating in the hand. The problem is most likely in the shoulder area, but could be closer to the spine. A slipped spinal disk or injury can result in stimulation of both the motor and sensory nerves in the area, causing both muscle spasms and tingling.

2. Problems with the reproductive organs in women often lead to complaints of lower back pain. Explain in terms of referred pain why this may be a common complaint.

Answer: Pain from the uterus is often "referred" to the lower back. This is due to poor localization of internal organ pain, and the referral of such pain to the surface area of the body whose sensory nerves enter at the same spinal segment as the organ's sensory nerves.

3. Peter has spilled acid on his hand. He said it burns, yet the acid was at room temperature. Explain in terms of stimulating a receptor with an alternate modality, how this sensation occurred.

Answer: Acid, due to its ability to damage tissue, may stimulate receptors or attached sensory nerves for both heat and pain. This combination is interpreted by the brain as "burning."

4. Jane's grandmother tends to hold papers very far away as she attempts to read them. She most likely has what visual disorder?

Answer: Presbyopia: stiffening of the lens and loss of the ability to accommodate to clearly see near objects.

5. A person who focuses light in front of the retina would have which visual disorder?

Answer: Myopia or near sightedness.

6. Which cones would be stimulated by the color purple? By the color aqua?

Answer: Purple: red and blue; aqua: blue and green.

7. Jill's closet light is burned out. The closet is only dimly lit from the room light. When Jill arrived at school, she realized she was wearing one blue shoe and one black shoe. Why did this happen?

Answer: Low light tends to poorly stimulate cones that are less light sensitive. Rods are more sensitive and are stimulated by low light, but do not determine color. The low light was ineffective in stimulating the cones well enough to clearly determine color.

8. Roger's grandfather has had a Weber's hearing test performed because he can't hear well out of his left ear. The tuning fork sound was louder in his right ear than his left. He asks you if a hearing aid might help him.

Answer: Grandfather has conduction deafness, since the tuning fork sound was louder in the affected left ear.

9. Action potentials on the cochlear nerve are arriving at the brain with a very high frequency, and are arriving primarily from nerves in the cochlea far from the oval window. What type of sound is this?

Answer: High frequency = loud
Far from the oval window = low pitch
Conclusion: A loud, low-pitched sound.

10. Uncle Joe has been diagnosed with an inner ear infection. He feels dizzy and nauseous and asks you to explain why he feels like the room is spinning. (Hint: Action potentials are generated in the vestibular nerve–how are these interpreted by the brain?)

Answer: A inner ear infection may cause action potentials to be generated in the vestibular system or the vestibular nerve. These are interpreted by the brain as originating from the normal stimuli of the vestibular system that is motion. Thus, he feels like the room is spinning.

11. During a high fever, patients complain of a "ringing in the ears." Explain this phenomenon in terms of the physiology of the sensory system and the brain's interpretation of action potentials on the cochlear nerve.

Answer: Action potentials in the cochlear nerve are interpreted by the brain as sound, even when the action potentials were created through other modalities.

Spinal Reflexes & Electromyogram (EMG)

1. How might an EMG be used to teach a patient to relax tense muscles?

Answer: In watching the amplitude of the EMG waves, a patient can attempt to reduce the amplitude. This can teach the patient to relax the muscles.

2. Tubocurare blocks the nicotinic receptor on the skeletal smooth muscle. This drug is shot into animals in Africa to paralyze them. Why would this drug paralyze the animals? In the presence of this drug, would their alpha motor nerves still have an action potential? Would their skeletal muscles still have action potentials?

Answer: The drug would paralyze the animals because an action potential in the alpha motor nerve could not be transmitted to the muscle without the nicotinic receptor to create the end plate potential.

The nerves would still have action potentials, but the skeletal muscle would not produce a large enough, if any, end plate potential to bring the membrane to threshold.

3. Myasthenia Gravis is a disease in which the nicotinic receptor on the skeletal muscle is destroyed by the patient's own immune system.
   a. What type of symptoms would this patient have?

Answer: Muscle weakness, poor motor control.

   b. Would this patient have a muscle action potential for every alpha motor neuron action potential? Why or why not?

Answer: Probably not. Fewer receptors would result in fewer open Na^+/K^+ channels in the motor end plate. Less Na1 entry into the end plate would result in a smaller end plate potential, often too small to reach threshold in the muscle cell.

   c. The treatment for this disease is to give the patient a drug to inhibit acetylcholinesterase. Why would this treatment be helpful?

Answer: This would effectively slow the breakdown of acetylcholine. Thus, the acetylcholine released with the alpha motor nerve action potential would hang around the motor end plate longer, and repeatedly bind to the available nicotinic receptors. Repeated opening of the few Na^+/K^+ channels available would increase the Na1 entry into the end plate, and increase the size of the end plate potential, improving the chances of bringing the muscle cell to threshold.

Electrocardiogram (ECG)

1. Angela Angina had a heart attack and died almost instantly. Attempts by paramedics to revive her were unsuccessful. What arrythmia did she most likely have?

Answer: Rapid death most often results from ventricular fibrillation. Blockage of a coronary artery inhibits blood flow to an area. This area does not depolarize and repolarize properly. Conduction must then go around the affected area increasing the chances of a circle rhythm and ventricular fibrillation. The ventricle does not pump effectively when fibrillating, resulting in brain death.

2. Rodney Rapid has been diagnosed with ventricular tachycardia. What is setting his rapid ventricular rate? What would his rate of atrial contractions be? What would you predict about his end diastolic volume or stroke volume?

Answer: Assuming no other abnormality, the ventricular rate is most often set by an ectopic foci located in the ventricle or lower portion of the atrioventricular (AV) node. Because the AV node does not readily conduct backward, atrial contraction rate would be set by the sinoatrial (SA) node and would be normal or slightly raised if the sympathetic nervous system is stimulated. His end diastolic volume (EDV) and stroke volume would likely be reduced if ventricular rates were greater than 200 beats per minute.

If atrial rate is also fast, the diagnosis is more complex and would include an ectopic atrial foci setting the rapid rate, atrial fibrillation, or a re-entry phenomenon in which the depolarization passes from the ventricle to the atria and back to the ventricle in a continuous circle.

3. During atrial flutter, the ventricle does not beat as rapidly as the atria? Why is this so? Is the AV node diseased in these patients?

Answer: Depolarizations arrive at the AV node when it is still in refraction from the previous depolarization. Thus, some depolarizations do not get the ventricle, and the ventricular contraction rate is slower. The AV node is not diseased, but is functioning normally.

Properties of Cardiac Muscle

1. Well-trained marathon runners have slower resting heart rates than untrained individuals. Predict the end diastolic volume in these athletes. Predict the stroke volume. Explain.

Answer: Assuming these runners have a normal resting cardiac output, and their sympathetic nervous system is not stimulated at rest (no increase in contractility), their EDV is likely greater and results in a larger stroke volume (Frank– Starling). Stroke volume must be higher if heart rate is lower and cardiac output is normal.

2. Caffeine causes extra Ca^{2+} to be released from the sarcoplasmic reticulum of the cardiac muscle cells in response to an action potential. Predict the effect of caffeine on the contraction strength of the ventricle. Explain.

Answer: Extra Ca^{2+} = greater Ca^{2+}-troponin binding, more uncovered actin binding sites, more cross-bridges formed and a stronger contraction.

Regulation of Mean Arterial Pressure

1. Debbie Dry has run a marathon, didn't drink enough water, and has become very dehydrated.
    a. What change would dehydration cause in the cardiovascular system?

Answer: Reduced blood volume = reduced venous return = reduced EDV, stroke volume, and cardiac output.

   b. Debbie's blood pressure is normal. How can this be so?

Answer: She has compensated by increasing her sympathetic stimulation:

venous vasoconstriction = returning venous return toward normal

increased contractility = returning stroke volume and cardiac output toward normal

increased heart rate = returning cardiac output toward normal

increased total peripheral resistance (TPR) = returning blood pressure toward normal.

   c. What would you predict about Debbie's heart rate? Her TPR? What about her pulse pressure?

Answer: Faster heart rate, higher TPR, and lower pulse pressure.

2. Artificial respiration is performed by blowing air into the patient's lungs. This is the opposite of a normal, "suction" inhalation. What affect might artificial respiration have on venous return?

Answer: Loss of the inhalation effect on venous return would reduce venous return.

3. Barry Bleed has been in an accident and lost almost 2 pints of blood.
    a. Predict Barry's venous return. (↓)

   b. Predict Barry's cardiac output. ($\downarrow$)

   c. When lying down, Barry has a normal MAP, even though his cardiac output is low. How can this be so?

Answer: A low cardiac output may be compensated for by a higher than normal TPR, to a certain level of cardiac output.

$$(MAP = CO \times TPR)$$

   a. When standing up, Barry's MAP is below normal.

   b. Predict Barry's heart rate when standing. (increased)

   c. Predict Barry's TPR when standing. (increased)

   d. Predict Barry's cardiac output when standing. (below normal)

Blood Cells

1. Puppies are commonly vaccinated several times when young because of immunity (maternal antibodies) transferred from the mother to the puppies during pregnancy and through the milk.

   a. Why would the maternal antibodies in the puppy prevent the puppy from producing immunity from the vaccine?

Answer: Antibody will bind to the antigens on the virus and cause virus destruction by the immune system before the virus has an opportunity to stimulate the appropriate B-cell.

   b. What kind of immunity do puppies get from their mother (active or passive)?

Answer: Passive immunity–they receive only antibody and do not generate memory cells.

   c. If a puppy has immunity obtained from the mother, why does he need to be vaccinated?

Answer: The antibodies eventually are destroyed by the body, leaving the puppy susceptible to infection.

2. If a person is exposed to influenza virus type 7, why isn't he immune to influenza virus type 6?

Answer: Unless the viruses have similar antigens, the memory cells will not respond to the type 6 virus.

3. A person while sick with one virus does not commonly get a second viral infection at the same time. If immunity is specific for one virus, what prevents the second virus from infecting the patient?

Answer: Nonspecific immunity is stimulated, and the activity and chemicals of this system can destroy a second invader before it gets well established.

4. One type of heart damage is believed to be caused by the body's immune response to a Strep bacteria. How can the immune system's response to a virus cause damage to a patient's own body?

Answer: The antigen on the virus is similar to the antigens on the myocardial muscle. The antibodies produced to destroy the virus also react to the heart cells, stimulating an attack of the immune system on the heart.

5. Jane Doe has a neutrophil count of 55% (normal = 55–75%). Her total white count is 2000 cells/mm^3 (normal = 5000–10,000). Does she have a normal <u>number</u> of neutrophils?

Answer: With a total white count of only 2000, her actual number of white blood cells is $0.55 \times 2000 = 1100$. This is below the normal of 5000.

Lung Volumes & Capacities

1. Josie thinks she may be having an asthma attack. What test would you use to determine if she has asthma? What change would you see in the test?

Answer: $FEV_{1.0}$. Percent should be below 60 for a diagnosis of asthma.

2. Bill has fallen on a sharp piece of metal that has cut an opening in his chest wall. What would happen to his intrapleural pressure? Is there any way you could help him breath?

Answer: Intrapleural pressure would no longer be below atmospheric, but equal to atmospheric. His lung would collapse on that side. Inflation of the lung could be accomplished by positive pressure ventilation–artificial respiration.

Regulation of Respiration

1. Calamity Jane has fallen off her bike and cracked several ribs. This has made respiration painful and she is not breathing well.
   a. What is this condition called?

Answer: Hypoventilation.

   b. Predict the following values when compared to a normal individual:

alveolar pCO_2 ($\uparrow$)
blood pCO_2 ($\uparrow$)
alveolar pO_2 ($\downarrow$)
blood pO_2($\downarrow$).

   c. What would you predict about her blood pH? Explain.

Answer: Should be low: high blood CO_2 results in the formation of carbonic acid.

   d. If she was given a higher oxygen content in her inhaled air, would this correct her blood pO_2? Her blood pCO_2? Her blood pH?

Answer: Her blood pO_2 would be corrected, but pCO_2 and her pH can only be corrected by increasing ventilation.

2. Carbon monoxide poisoning is treated with inhaled air that is high in oxygen and also has CO_2 added. Why might this be beneficial in eliminating carbon monoxide from the body?

Answer: Elevating arterial pCO_2 would stimulate the respiratory center, increase ventilation, and increase the rate of elimination of carbon monoxide from the body.

3. Randy Renal has a kidney infection. He is unable to eliminate all the acid from his body, and he has a metabolic acidosis–a blood pH that is low (high level of H^+ in the blood).
   a. What would you predict about his respiratory rate and depth? Explain.

Answer: Increased H^+ would stimulate respiration, regardless of the origin of the H^+.

   b. What would you predict about his blood pCO2? Explain.

Answer: pCO_2 would be reduced because of the increased ventilation.

Kidney/Urinalysis

1. Joe has had the stomach flu for 2 days and is very dehydrated. Predict Joe's ADH level (high, low, or normal). Predict Joe's urine volume and specific gravity (high, low, or normal).

Answer: ADH: high

Answer: Urine volume: low with high specific gravity.

2. Billie has a urine specific gravity of 1.008, yet he is dehydrated. What hormone might Billie be unable to produce?

Answer: ADH.